DOMESTIC
SOLAR
ENERGY

A Guide for the Home Owner

DOMESTIC SOLAR ENERGY

A Guide for the Home Owner

GAVIN D. J. HARPER

THE CROWOOD PRESS

First published in 2009 by
The Crowood Press Ltd
Ramsbury, Marlborough
Wiltshire SN8 2HR

www.crowood.com

British Library Cataloguing-in-Publication Data
A catalogue record for this book is available from the British Library.

ISBN 978 1 84797 060 2

Disclaimer
The author and publisher do not accept any responsibility, in any manner whatsoever, for any
error, or omission, nor any loss, damage, injury, adverse outcome or liability of any kind
incurred as a result of the use of any of the information contained within this book, or reliance
upon it. Readers are advised to seek specific professional advice relating to their particular
property, project and circumstances before undertaking any installation work concerning
domestic solar energy.

Frontispiece image courtesy Solar UK.

Typeset by Jean Cussons Typesetting, Diss, Norfolk

Printed and bound in Malaysia by Times Offset (M) Sdn Bhd

CONTENTS

FOREWORD

The prospects for solar energy are looking bright. Governments around the world have at last begun to recognize the enormous potential of this clean, inexhaustible energy source and are implementing policies to encourage its development and deployment on a very large scale.

This will involve a major expansion of the already substantial world-wide solar manufacturing industry, supported by a growing network of designers, consultants, retailers and installers, backed up by an appropriate system of legislation and regulation. But it will also need a constituency of well-informed users, able to make wise choices about the types of solar energy system that are best suited to their individual circumstances and requirements.

This book will do much to inform that constituency. Aimed at householders keen to install solar energy systems in their own homes, it starts with explanations of the principles underlying the use of solar energy, described in non-technical language with admirable clarity.

It then describes the 'passive' use of solar energy in the design of houses, to enable the sun to contribute to their space heating require-ments. Subsequent chapters explain, clearly and simply, how solar energy can contribute to a home's interior lighting, power its electrical appliances and supply most of its water heating needs. Householders will, of course, want to know how long it will take to recoup the capital investment in a solar installation through savings in energy bills. They will also need to be aware of legislation that could affect their solar installation. These topics are thoroughly covered in the final chapters, augmented by helpful appendices describing relevant solar industry bodies and standards, giving 'sun-path diagrams' to assist with the orientation of UK solar energy systems, and listing useful addresses and further reading on the subject.

In short, if you want to know more about how solar energy can reduce your home energy bills – and cut your carbon footprint – this excellent book should give you most of the information you need.

Godfrey Boyle
Director, Energy & Environment Research Unit
The Open University
December 2008

DEDICATION

Every once in a while, someone really special comes along who gives you immense confidence in your abilities, and helps you to realize your dreams – to Judy Bass, for giving me my first chance to write, the encouragement to keep writing and making me laugh in between.

ACKNOWLEDGEMENTS

I am indebted to a great number of people for their help and production of this book, all of whom have been exceedingly generous in sharing their knowledge, experience and insight, which has fed into the production of this book.

I would like to express my gratitude to Alison Pooley, as without her introduction to The Crowood Press, this book would never have come into being. I am also indebted to Alison for giving me the confidence to draw.

I am also continually inspired and encouraged by the students on the MSc Architecture: Advanced Environmental & Energy Studies Course, year of 2005–06. A particular thank you to Maria Hawton-Mead, who has provided insight on domestic solar energy best practice in Germany and helped form my ideas for the sections on solar shading.

A big thank you also to Phil Neve and also to Brandon Oram of Mr Pym the Plumber for his counsel on solar installations.

Thanks are also due to Ben Robison of Dulas Ltd, for the generosity of his time in helping me to source images and for sharing his insight on Dulas' work.

I'd like to convey my thanks to Tim Godwin and Oliver Sylvester-Bradley at Solar Century, a truly innovative firm that continues to be at the cutting edge of solar-energy innovation in the UK.

Thanks also to Rob Yarrow of Sol-Heat, and to Rebecca Ashley at Schott UK, for helping to source images and information on Schott's evacuated tube collectors.

Thanks also to Edel Walsh for kindly allowing me to use Invisible Heating's images.

Also, a big thank you to Nikki Wright of SolarUK – installers and manufacturers of solar equipment – for her help in locating images of SolarUK's products.

INTRODUCTION

In this book we are going to be examining what you, as a home owner or builder, can do to your house to make optimum use of solar energy. I must give you hearty congratulations for buying this book, not for the royalty that I'll earn, but for being 'part of the change'. By showing an interest, you've made a positive step in the right direction to making the world a better place. The world needs more people like you.

The fact you're still reading, means that you've let me stick my foot in the door. In the pages of this book, I want to sell you a vision of the technologies that are going to deliver a brighter future, and I hope that by the time you've finished turning the last page, you will have bought into the dream.

I'm going to start with the big picture... the really big picture, and start 'zooming in' bit by bit. We're going to start by looking at the world and, in a number of short steps, we're going to end up looking at your home – a bit like typing your postcode into Google Earth.

The World

We, the human race, are collectively faced by a number of tough choices in the years ahead. The illusion that the world is infinitely vast has been shattered, with cheap transportation and the communications revolution. Technologies allow us to extract and utilize the earth's resources; however, as we begin to consume on an ever larger scale, it has become apparent that the finite resources of the world will only go so far.

Specifically, after a flurry of excitement we've gone from mass consumption of coal, to oil to gas (and then on to uranium), but we are fast coming to the realization that these are finite resources – leaving serious questions about where we are going to get our energy from in the future.

Marion King Hubbert was an American oil geologist who presented a paper to the American Petroleum Institute, on a theory which has now come to be known as 'Peak Oil'. He highlighted that exploitation of natural resources follows a symmetrical bell-shaped curve. With many of our natural resources, the prevailing notion is that we have climbed the steep hill and we are now standing at the precipice staring down the other side of the curve – with a grim future of resource shortage looming if we don't act soon.

There is also another glaring problem. Whilst we in the 'developed world' are already producing enough carbon dioxide to irreversibly change our climate, there are also other countries with bold ambitions to develop their own economies. This is only going to happen with massive amounts of carbon dioxide emissions or innovative solutions. It is hard to assume a moral high-ground and say 'don't produce carbon dioxide' when those same dirty technologies fuelled our own industrial revolution and development.

Europe

The European Union has set a target of producing 20 per cent of its energy sustainably by 2020; at the moment, certainly, the UK is not pulling its weight to meet this target. The issue of carbon dioxide emissions is going to be a contentious one over the coming years and, with growing consensus in the EU about emissions trading schemes, the countries that do not adapt to a low-carbon future will be left behind, paying for the privilege of sticking with outmoded, dirty technologies.

At the moment, we're used to cheap energy. UK energy prices are below the EU average.

The UK

Consider your investment in solar energy technology, as buying your energy in advance. We all seem to have warmed to the idea of pay-as-you-go phones, where you buy a voucher for calls in advance and then use them as you go. Well it's the same with solar energy: once the initial purchase has been made, the system will continue to return solar energy over the lifetime of the system.

For us in the UK, we are in the middle of a fundamental change in the way our nation sources its energy. Following the demise of the coal industry in the UK, our nation 'dashed' for the new-found oil and natural gas in the North Sea. We built lots of gas-fired power stations and have been happy for several decades enjoying the bounty of our own North Sea oil... however, the winds of change have blown. The UK is in the process of changing from becoming a net energy exporter, to a net energy importer.

As the gas and oil in the North Sea begin to decline, we are faced with the challenge of having to import increasing amounts of energy from abroad. This is coupled with the fact that Britain's ageing (some would say decrepit) fleet of nuclear reactors is on its way out, one by one reaching the end of their life – having long forgotten promises of delivering energy too cheap to meter.

There is much work to be done in the UK. The current government mechanisms for funding renewable energy development in the UK (especially for homeowners) leave much to be desired. Net metering schemes currently in place run a poor second place to 'REFIT' (Renewable Energy Feed In Tariffs), which have encouraged such aggressive development of solar energy resources in Germany. Britain looks like it will fail in meeting its share of the EU 20 per cent by 2020 target.

With the rise in costs for domestic energy – heating fuels and electricity all steadily increasing – homeowners are likely to be increasingly vigilant about the energy costs of running a home. Investing in optimizing your home's use of solar energy is the 'right' thing to be doing in an age of acceptance of anthropogenic climate change and concern about impending shortage of fossil fuels.

Your Local Authority

Innovative local authorities like the London Borough of Woking, have formulated innovative responses to the need for clean energy by creating local private wire networks to share renewable energy generated locally, without losing the value of the electricity by selling it to the grid at bargain price.

Merton has instituted a rule whereby planning applications for certain builds will require 10 per cent of the installation's energy to be provided by low-carbon methods. This trend towards renewable energy being integrated in planning applications is something that local authorities throughout the UK are beginning to wake up to.

Your Street

It is undeniable that, firmly rooted within the British psyche, is a culture of 'keeping up with the Jones''. This phenomenon can be observed all over the UK – one neighbour gets a new driveway and very soon it springs up all the way down the street. Similarly, your home can be a tool for promoting sustainable energy in your locality, and it is quite likely that once you install solar devices, you will cultivate interest amongst your neighbours!

Your House

With the introduction of Home Information Packs and Energy Performance Certificates for homes, buyers will be looking ever more closely at how much energy a home consumes – the energy performance of a house may be a crucial factor in selling a house in the near future, with rising energy prices.

You

Aside from reducing your energy bills in the long run, installing solar technology in your home should give you an enormous sense of pride and well-being, knowing that you are being part of the change, moving the world forward to an age of greater energy security through your actions and helping to reduce your personal impact on the planet.

Note

Most of the installations described in this book are not suitable for amateur 'do-it-yourself' installation. Having chosen the type of installation you wish to use, you should scrupulously follow the manufacturer's guidance on having it fitted. Seek professional advice and ensure that you are following all of the relevant regulations on the installation of electrical equipment and services.

CHAPTER 1

THE SOLAR RESOURCE

When faced with a wide array of technical products geared up for the homeowner who is interested in solar energy, it may seem an exciting prospect to jump straight into the technology, comparing products and gadgets on offer for your home. However, you first need to understand the basics of solar energy and the solar resource, before even beginning to look at how you might be able to utilize it in your home. This chapter should serve as a primer on solar energy.

To give you an idea of the enormity of all this, consider the fact that only about a billionth of the radiation from the Sun actually reaches the Earth.

The Sun

About 150 million kilometres away from the Earth (around 93 million miles) is a medium-size star. I use 'medium' in the scientific sense of the word, because when you consider that it accounts for 99.8 per cent of our solar system's mass, weighs the equivalent of 332,946 planet Earths, has the same volume as 1,300,000 Earths and has the surface area of 11,900 Earths… medium seems somewhat an inadequate adjective. In short – it's pretty huge.

Stars are classified by 'spectral classification'; our Sun is a 'G2V' star, where 'G2' indicates that the surface temperature is around 5,780° Kelvin (around 5,507° Celsius), which gives it a white colour. Our Sun doesn't just exist as a point in the sky – even though it is so far away – it occupies an area in the sky of about 32 minutes of arc, which translates into 0.533 degrees.

Our Sun is about 1.4 million km in diameter, which is about 109 times the Earth's diameter. To give you some relative size of scale, find a ball-bearing that is an eighth of a centimetre, and take a balloon and inflate it to 13.5cm in diameter. Held with their centres at a distance of 1.5m, this represents the Earth and Sun to scale.

Solar radiation is our primary source of energy on Earth. At the outer boundary of our atmosphere, an area $1m^2$ receives on average 1,367W of solar energy. The UK receives as much as 60 per cent of the energy that a similar landmass located on the Equator would receive, with this amount of energy being equivalent to well over 1,000 power stations.

The relationship between the Sun and Earth isn't a static one. Our Earth is constantly spinning around its axis, whilst rotating around the Sun. The spinning motion gives us our days, as points on the Earth face the Sun, and face away from the Sun, whilst the orbital motion around the Sun gives us our years and seasons.

An 'Earth revolution' is the time taken for the earth to complete a journey around the Sun. It takes 365.26 days to accomplish this feat – the extra 0.26 days gives rise to leap years to compensate for this extra fraction of a day.

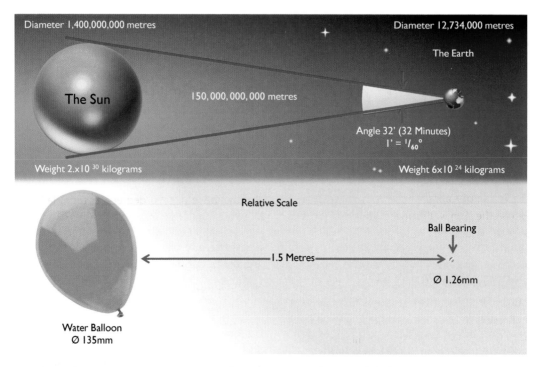

Diameter 1,400,000,000 metres   Diameter 12,734,000 metres

The Earth

The Sun   150,000,000,000 metres

Angle 32' (32 Minutes)
$1' = {}^{1}/_{60}{}^{\circ}$

Weight $2.\times10^{30}$ kilograms   Weight 6×10^{24} kilograms

Relative Scale

Ball Bearing

1.5 Metres

Ø 1.26mm

Water Balloon
Ø 135mm

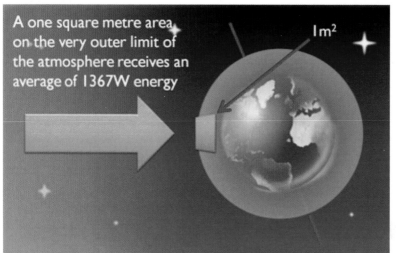

A one square metre area, on the very outer limit of the atmosphere receives an average of 1367W energy   1m²

TOP Relative sizes and distance of Sun and Earth.

LEFT Solar radiation falling on the atmosphere.

In fact, this isn't quite the whole story, as our Earth actually follows an elliptical orbit, which means that it's not a fixed distance away from the Sun. The 'Perihelion' in the Earth's orbit (3 January) marks the point at which it is nearest to the Sun, around 147,300,000km away, whilst the 'Aphelion' (4 July) marks the point at which the Earth is farthest away from the Sun at 153,100,000km. However, surprisingly, it is not this phenomenon that causes our seasons.

The Earth's axis runs through the Earth in an imaginary line from North to South Pole. The Equator is a line around the Earth, which runs perpendicular to the Earth's axis and 'cuts' the earth into two halves at its widest point. If the Earth spun on its axis vertically, our days would be the same length all the year around and we would not have any seasons. Instead, our Earth is inclined by around 23.5 degrees.

If we look on a globe, the Tropic of Cancer and the Tropic of Capricorn mark the boundaries of where the Sun will appear 'directly overhead' over the course of a year.

Summer Solstice – 22 June

On this day, every year, the North Pole will be inclined at an angle of 23.5 degrees towards the Sun. For us in the Northern Hemisphere, this is good news – it is the longest day of the year, during which we receive the most sunlight. At noon on the Summer Solstice, the sun will be

> **CELESTIAL DATES FOR YOUR DIARY**
>
> Perihelion – 3 January.
> Vernal/Spring Equinox – 20/21 March.
> Summer Solstice – 21/22 June.
> Aphelion – 4 July.
> Autumnal Equinox – 23/24 September.

directly above (i.e. 90 degrees altitude) latitude 23.5 degrees – the Tropic of Cancer. If you are above 66.5 degrees latitude, you will have 24-h sunlight. Although this doesn't apply to anyone living in the UK, it's good news for Shetland Islanders, as it doesn't get dark at night, it just gets a little less light (known locally as the Summer Dim.)

Autumnal Equinox – 23 September

On this day, the North and South Poles are equi-

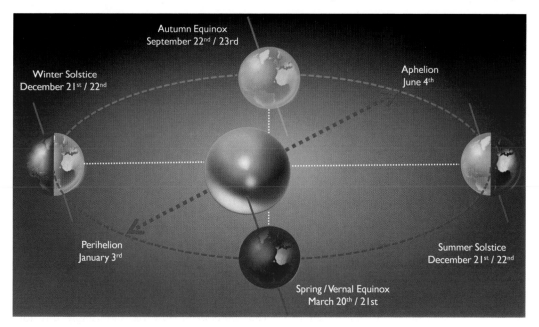

The relationship between the Sun and Earth.

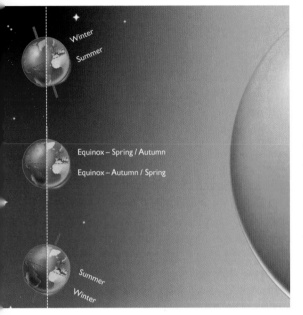

Winter
Summer

Equinox – Spring / Autumn

Equinox – Autumn / Spring

Summer

Winter

distant from the Sun – wherever you are on Earth, you will receive 12 hours of sunlight and 12 hours of darkness. Everywhere on Earth, the Sun will rise at 06.00 in the East and set at 18.00 in the West.

Winter Solstice – around 23 December

This is the opposite to the Summer Solstice – the South Pole is now inclined at an angle of 23.5 degrees towards the Sun. In the Northern Hemisphere, the news isn't so bright for solar fans – it is the shortest day of the year, when we receive the least sunlight.

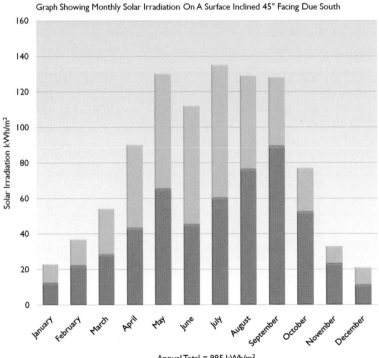

Graph Showing Monthly Solar Irradiation On A Surface Inclined 45° Facing Due South

Solar Irradiation kWh/m²

Annual Total = 985 kWh/m²
Direct Solar Radiation = 537kWh/m²   Diffuse Solar Radiation = 448 kWh/m²

■ Direct Solar Radiation   ■ Diffuse Solar Radiation

TOP LEFT **How solar radiation hits the Earth during different seasons.**

BOTTOM LEFT **Monthly solar irradiation on a surface inclined 45 degrees facing due south.**

Vernal Equinox – 21 March

This is also known as the Spring Equinox. By this time, the Earth is in the opposite position (on the other side of the Sun) to where it was at the Autumnal Equinox. The Sun is therefore equidistant from both poles and so we will receive 12 hours of sunlight and 12 hours of darkness. Everywhere on Earth, the Sun will rise at 06.00 in the east and set at 18.00 in the west.

However, the total amount of solar radiation we will receive at our property depends on much more than simply the latitude of our site. Local weather conditions, affected by the geography of our site and its environs, also impact the amount of 'direct' solar radiation that falls on our site.

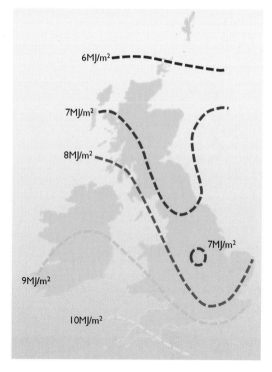

UK March average daily total solar energy in MJ/m^2.

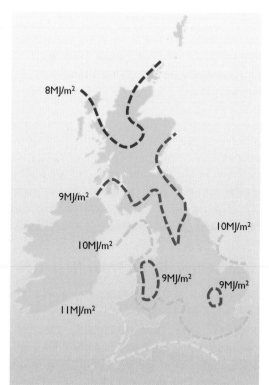

UK annual average daily total solar energy in MegaJoules (MJ) per m^2.

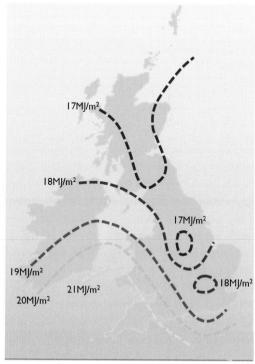

UK June Average Daily Total solar energy in MJ/m^2.

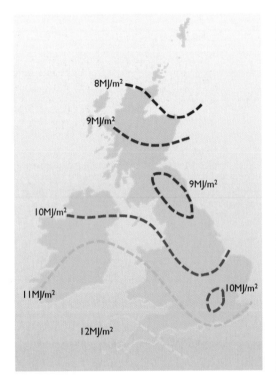

UK September average daily total solar energy in MJ/m².

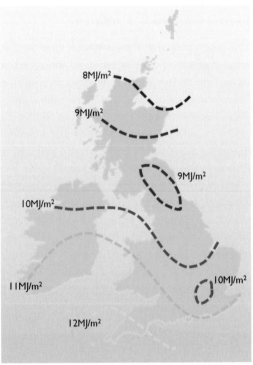

UK December average daily total solar energy in MJ/m².

Unfortunately, when we begin to assess our heating requirement for energy, and match it up to the available solar energy, we find that there isn't much correlation – most of our requirement for heating comes in the winter, which is precisely when the amount of solar radiation reaching a horizontal surface is at its lowest. However, the solar energy that is available can still be profitably used for domestic hot-water heating and for the generation of electricity.

With climate change a subject that is finally receiving the attention it deserves, there is probably going to be an increased focus on how we can adapt our built environment to changing weather conditions in the UK. It is clear to many, that our summers are getting hotter –

as evidenced by the growth in sales of portable air-conditioning units into the home market of late, and the news of vineyards and olive groves in isolated spots in the south of England.

Availability of the Solar Resource

Although our demand for space heating is the highest in the winter, when our solar resource is at its lowest, our demand for domestic hot-water remains relatively constant throughout the year. The graph opposite shows the relationship between our demand for solar energy and its availability over the course of the year.

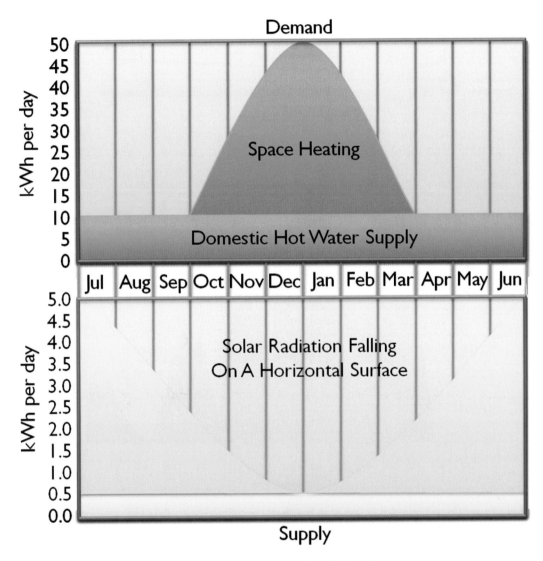

The relationship between our heating energy needs and available solar energy.

The Nature of the Sun's Light

The Sun's Energy, Frequency and Wavelength

We see light in the range of wavelengths we term 'visible light'. However, visible light constitutes just a part of the electromagnetic spectrum – infra-red and ultra-violet light also make up a significant fraction of the light energy that reaches the Earth from the Sun. The image on p.18 shows the spectral distribution of solar radiation at the Earth's surface.

Different surfaces absorb and reflect light differently. A black surface will absorb light hitting

17

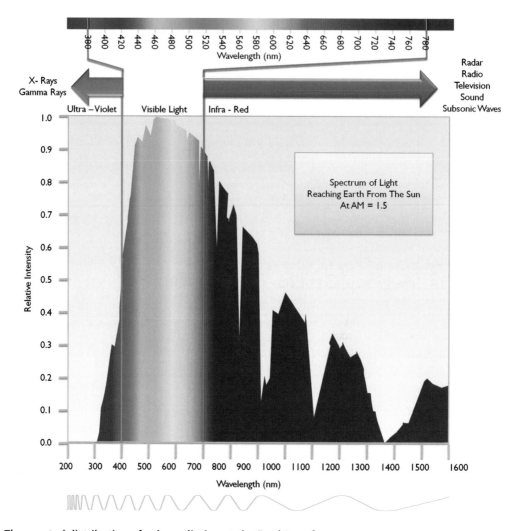

The spectral distribution of solar radiation at the Earth's surface.

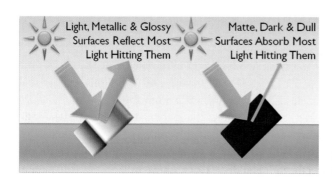

Different coloured surfaces reflect and absorb light differently.

it and heat up, whilst a stainless steel polished mirror will reflect light. Accordingly, we colour 'collector' surfaces black or use a special coating called a 'selective surface' to harness the most solar energy.

Air Mass

As the Sun changes position, the amount of 'atmosphere' that the light has to travel through changes dramatically. AM 0 refers to the solar spectrum that is measured from outside of the atmosphere. AM 1.0 refers to the solar spectrum once it has travelled through one thickness of atmosphere (measured perpendicularly from the Earth's surface). However, as sunlight travels through the atmosphere obliquely, the amount of atmosphere the light must travel through changes – if the light travels through a thickness of one and a half atmospheres, we denote this as AM 1.5.

Indirect and Direct Solar Radiation

Changing the orientation of our solar collector influences the amount of direct and indirect radiation it receives in different ways. By tilting the collector towards the sun, we increase the amount of 'direct' solar radiation that it receives; however, we reduce the collector's exposure to the sky hemisphere, which reduces the amount of indirect radiation the collector receives.

Solar Orientation

In this book we are going to discuss a variety of different technologies that can be used to capture and harness solar energy – whether for producing thermal or electrical energy. All solar devices need to be positioned correctly, in

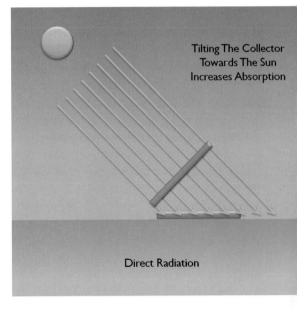

Direct radiation – tilting towards the sun maximizes absorption.

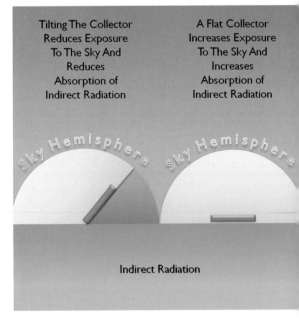

Indirect radiation – tilting from the horizontal minimizes exposure to the sky and decreases absorption.

19

order to make the best use of the solar resource; however, decisions on how to site devices are not always easy or straightforward.

Some methods of mounting allow for adjustment or flexibility of the orientation of the solar panel – in this extreme, this is embodied in the solar tracker, which can follow the Sun throughout the day, automatically ensuring that the cells are facing the Sun directly and, therefore, capturing the best portion of the solar energy. However, mounting systems like this introduce additional cost and complexity, so the benefits must be carefully assessed. Some compromise may be necessary – a slightly more inflexible, but simpler and cheaper system, is to have the cells fixed in azimuth, but with a variable angle to the horizontal, allowing the altitude of the sun to be tracked by adjustment over the seasons. Of course, if your solar panels are integrated into your roof, or the façade of a building, you may not have any control over the angle or orientation of your panels, necessitating a static, fixed installation.

Angle from the horizontal:

■ At the equinoxes, if the angle of tilt of the solar collector from the horizontal equals the latitude of the site, the collector will be perpendicular to the Sun's rays.

■ Changing the collector angle can optimize the collector to the Sun's changing position over the seasons.

■ To improve performance, the angle of the collector can be increased by 15 degrees in the summer and decreased by 15 degrees in the winter.

■ As we deviate from due south, the amount of energy that is captured decreases.

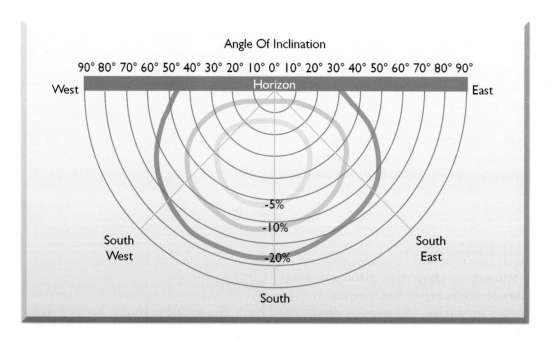

This graph shows the effect of altitude and azimuth deviation on solar gain.

Over the period of the course of a year, the proportion of direct and indirect radiation changes, as can be seen in the graph opposite.

If we look to the image below, we can gauge approximately how much energy we will lose by deviating from the optimum altitude and azimuth orientation.

Sun-Path Diagrams

Sun-path diagrams are a useful tool to enable you to gauge the position of the Sun in the sky, at a given latitude, at a given time of year and at a given time of day. Different Sun-path diagrams are produced for different latitudes – we have supplied the most commonly used Sun-path diagrams for the UK here in this book.

How to Read a Sun-Path Diagram

First of all, we are going to pay attention to the horizontal and vertical lines on a Sun-path diagram. Imagine a semi-transparent sheet of squared paper, which you hold directly in front of your face. Face south and curl the paper around your field of vision. In your peripheral vision, you can see east to the left of you, and west to the right of you. If you stood in the same place for the duration of the day, from sunrise to sunset, you would see the Sun plot a curve through the sky. Imagine a pin-hole camera making a day-long exposure of the scene – a white streak across the picture, which forms a bump like curve. Now, relate this curve to the squared sheet of paper – as the world turns, the Sun will make its path across your field of vision, slowly crossing the lines on the squared paper.

However, the horizontal component of the Sun's motion is only part of the story. The Sun appears to start low in the sky at dawn, rises through the sky until midday, when the Sun is at its highest point, and then descends again until sunset, when the Sun is at its lowest point as it disappears below the horizon.

Looking at the image (*see* p.22), we can see that by combining these two components, we get the Sun's path – shown in the orange lines in the bottom graph. The different lines represent different times of the year, whilst the red lines represent times of the day.

Solar Energy by Any Other Name...

It is easy to take a very narrow-minded perspective of solar energy, and restrict your thoughts to photovoltaic solar panels, solar thermal collectors and other technologies, which take solar energy directly and transform it into something useful that we can use. However, looking at other renewable-energy technologies, we can see that all the natural process that provide their raw energy source, can all be seen to derive from the Sun – and so, with the exception of tidal power (which comes from the tides caused by the Moon orbiting the Earth), all renewable energy can be thought of as 'solar energy' in one form or another.

Although this book is not the place for a thorough exploration of all renewable-energy technologies, these technologies should be noted, as they are not mutually exclusive, but complementary. Renewable-energy sources are intermittent, to a degree, however, including a portfolio of renewable-energy sources gives us a measure of protection – when the Sun is not shining, the wind may be blowing – and so on and so forth.

Biomass

Plants take in carbon dioxide from the atmosphere, water from the ground and other food and nutrients. The leaves of plants are like solar

The horizontal lines on the graph give the **Altitude** of the Sun relative to the horizontal ground plane.

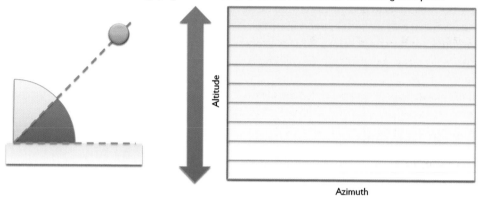

The vertical lines on the graph give the **Azimuth** of the Sun relative to the horizontal ground plane.

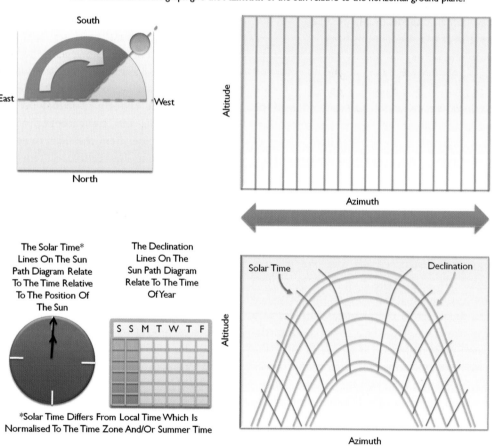

How to read altitude, azimuth, solar time and declination angles on a Sun-path diagram.

collectors – the Sun's energy allows a process called photosynthesis to occur, where the plants take the Sun's energy and use it to convert the carbon dioxide, water and nutrients into energy, which can be used by the plant to enable it to grow. Once the plant has grown, we can cut it down, burn it and produce heat in the process. So, as a plant grows, it is effectively sequestering solar energy in its cells, ready for later use.

In sustainably managed biomass, as one lot of plant matter is harvested, new plants are planted to ensure that the biomass is always growing, and plant matter is only taken as fast as it can grow. Now, burning plant matter clearly produces carbon dioxide; however, this is carbon dioxide that has already been taken out of the atmosphere during the plant's life – so the only carbon emissions we need really to consider are those produced as a result of producing the biomass from agricultural machinery, processing, transport and so on. In your home, a biomass boiler can provide a back-up, enabling you to produce domestic hot water, on days when the Sun isn't shining. You can also use biomass for space-heating.

Wind Power

The Sun heats up the air in our atmosphere. As air is heated, it becomes less dense. The changes in the local density of pockets of air cause the air in our atmosphere to circulate as a result of a variety of different mechanisms.

Hydro Power

The Sun drives the hydrological cycle, the cycle whereby water evaporates from the Earth's surface, is held in the clouds and then rains back down to earth. Where water lands on higher ground, it flows back to lower ground under the influence of gravity. We can harness the energy in falling water to generate electricity.

CHAPTER 2

PASSIVE SOLAR DESIGN

Designing for Efficiency

Before considering how to heat your house using solar power, you must carefully evaluate the energy performance of your building.

Our buildings are constantly losing heat to their environment as a result of various heat losses, and gaining heat from the Sun and from passive gains inside the house (which is the heat emitted by objects such as people, appliances and other devices that consume energy in the home and produce heat). Traditionally, we build buildings that require some additional energy in order to maintain a comfortable temperature –

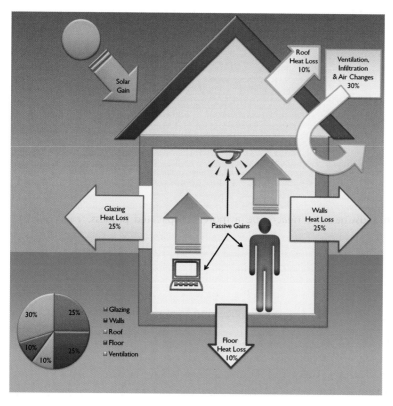

Heat losses and gains in a passive solar house.

in the winter, they lose too much heat, so we require an input of energy in order to raise the internal temperature; whilst in the hot summer sun, they gain too much heat – and there is an increasing trend to buy air conditioning, which requires yet more energy to 'extract heat' from the house.

The optimum situation is a house that requires little in the way of energy input to either heat or cool the house. Passive gains and solar gains from the Sun provide enough energy to keep the internal temperature comfortable in the winter, whilst in the summer, appropriate shading protects the house from the hot summer Sun and overheating.

In the figure opposite, we can see some of the heat losses in a house of traditional construction. Note that for clarity, no heating or cooling system is included in this diagram.

We can see that the bulk of the heat is lost from the house by ventilation, infiltration and air changes, which account for 30 per cent of heat loss in an average building. This is usually as a result of poor building design. A quality building structure will have close-fitting doors and windows, which do not allow any draughts to enter the building, sound brickwork and joints between items of the building fabric, which do not allow for air to escape; also smart occupant behaviour is important, which prevents heat from being lost. How many times have you seen someone in a house that has just had the 'heating turned on' later open a window to cool down? Through this action 'energy' is being used to heat the street – energy that you, as a homeowner, must pay for.

Next, 70 per cent of heat is lost through the fabric of the building, with this being apportioned around 10 per cent through the roof and floor each, and around 25 per cent being lost through both the walls and windows. There are a number of steps we can take to reduce these

HOORAY FOR HIPS

The Government has introduced Home Information Packs (HIPs), which will include, as part of their information, Energy Performance Certificates. The Energy Performance Certificate is a good place to start when looking at how energy is used in your home. It should give you a good guide to where you might be able to use energy more efficiently.

losses. Installation of high-quality triple-glazed windows can help to reduce the amount of heat lost through windows – which, considering their small size, are responsible for an awful lot of heat loss. Furthermore, insulation can be retrofitted to walls – there are a number of methods of achieving this: internal insulation, external insulation and cladding; or, if your property is of a cavity-wall construction, insulation can be sprayed into the void space between walls. If, on the other hand, you are looking at a new-build property, you have the luxury of considering at an early stage how to maximize insulation and minimize heat loss through design.

The roof space is a relatively easy project to tackle insulating – unless, of course, your loft has been converted to a room, in which case it can be a more complicated process. The floor can present some challenges to improving its performance on an existing property.

The Solar Resource and Space Heating

One of the reasons that we have neglected to consider the solar resource as a good way to heat our homes, is its particularly poor timing. When you consider your home's demand for heat, and compare it against the availability of solar energy, you will realize that whilst there is a lot of Sun in the summer, you need more heat in the winter. Furthermore, the challenge of how to heat your home using solar energy is exacerbated by shorter term variations – the Sun is there in the day, but not at night. Furthermore, there are often clouds in the sky during the colder months, when what we really want is the Sun. None of these problems is insurmountable – however, they do require a little bit of thought to make the best use of the direct solar energy that is available.

As a result of these challenges, there are often many compromises when taking an existing home and optimizing it for passive solar space heating. However, if we have the luxury of designing a new house and considering passive solar design from an early stage, it is possible to integrate passive solar principles into the fabric of the house – as passive solar design can greatly influences the form and design of the final building.

Passive Solar Design Principles

Correct Orientation of Your Building

Use a compass to find where is south on your site – consider facing the glazed areas or your house towards the south, south-east and south-west, to ensure that they receive solar gain during the heating season.

Look at the landscape and the form of the

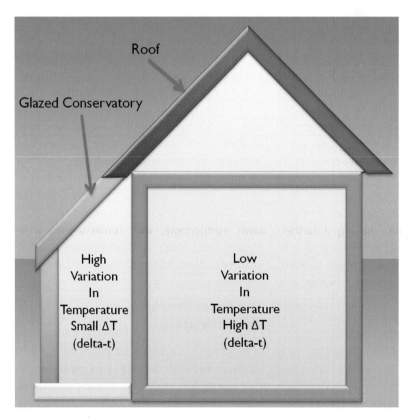

Adding a conservatory can provide solar gain and a buffer zone.

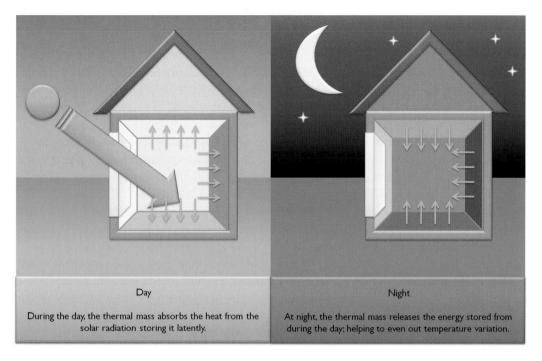

Day	Night
During the day, the thermal mass absorbs the heat from the solar radiation storing it latently.	At night, the thermal mass releases the energy stored from during the day; helping to even out temperature variation.

Thermal mass absorbs heat in the day and releases it at night, regulating temperature.

land, to determine whether there are any features, hills, mountains or slopes that would obscure the low winter Sun during the summer months. Look also for any tall trees or other obstructions that would obscure your building's access to the Sun. In some cases, where obstructions will severely limit your building's access to sunlight during the winter, consider re-siting your building on the plot, moving it further away from the obstructions – or, if the option is available, consider removing the obstructions, by judiciously pruning/cutting down trees, or by re-landscaping.

Passive Solar Design Elements

Conservatories

If you have a south-facing garden, you could consider a conservatory as an addition to your home. In the summer, this will provide an abundance of warm air, which can be trapped in the conservatory and let into the house to keep it warm as and when required. At evening, the conservatory must be sufficiently insulated from the main building fabric, so that heat will not escape at night when the conservatory cools down. Furthermore, with conservatories, it is important to provide appropriate solar shading and adequate ventilation, to ensure that excess heat does not cause a problem.

Thermal Mass

Thermal mass is an important tool in the Passive Solar Designer's toolbox. Thermal mass can be used in a variety of ways to 'capture and store' heat at times when it is in abundance, releasing it at times when the rest of the building cools. This provides the effect of evening out fluctuations in temperature: a building structure that is 'thermally massive' will tend to have a steady temperature. Consider the variations of temperature in a summer house or shed, compared to, a building of masonry construction. The temperature in the shed will vary with the seasons, heating in the summer to the point of excess during the day, and chilling to the point of freezing at night. There is nothing in the light structure of the shed that can 'retain' any heat, whereas the temperature in the solid masonry building may tend to be more stable as the walls, ceiling and floor absorb heat during the day and re-radiate it at night.

Thermal mass can be strategically positioned in Passive Solar Design, in such a way that light will fall on it and heat it – the colour and texture of the surface has a bearing on how well the thermal mass can absorb the solar radiation. Light surfaces reflect radiation, whilst dark surfaces absorb radiation.

Take the example of the building shown on page 27. Let's assume that the building is of a heavyweight construction, using materials that are 'thermally massive'. During the daytime, as the sun shines through the room, the building fabric absorbs solar energy, which in the adjacent picture, is then re-radiated when night falls.

Water is a very effective form of thermal mass, however, unfortunately because of its fluid nature, it cannot be used to form a structural element. Masonry, concrete, or alternative materials such as 'rammed earth' can form elements of thermal mass within a building.

This is a basic explanation of the principles of solar mass, however, to make the most of thermal mass in your building, some further reading and study of building physics will be required.

Solar Shading

One of the problems with introducing large areas of south-facing glass to a design, is that whilst it effectively captures solar energy when it is cold, it also captures an abundance of solar energy when it is hot. In some circumstances, this can lead to undesirable overheating.

This problem can be solved, in part, by solar shading, as part of a holistic design. Shading can be fixed, in which case the geometry of the shade is such that the low winter Sun can penetrate the building, whilst the high summer Sun is blocked from entry. Looking at the figure opposite, it is possible to see how something as simple as an overhang on a building can be used effectively to prevent the building overheating in the summer.

Additionally, shading can take other forms such as foliage or strategically planted trees. Shading can also be an integral feature of the landscape – positioning a new-build house relative to its surroundings so that appropriate shading is provided by natural features.

Additionally, there are many ways of creating functional moveable shades that are either operated manually, or by some form of actuator, to ensure that the building is shaded from excessive solar gain at times when the building is liable to overheat. Whilst a moveable shade entails an extra degree of complexity, it does lend flexibility to when and how the shade is operated, allowing occupants more control over the solar gain, and hence the building's temperature.

Trombe Walls

Trombe walls take their name after the French engineer Felix Trombe, who designed a simple method to harness solar energy using a blackened south-facing wall with an air-gap and

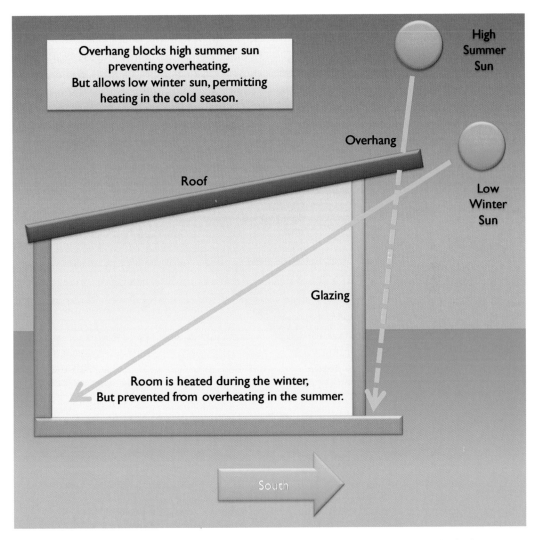

Overhang blocks high summer sun preventing overheating, But allows low winter sun, permitting heating in the cold season.

High Summer Sun

Overhang

Roof

Low Winter Sun

Glazing

Room is heated during the winter, But prevented from overheating in the summer.

South

An overhang permits the low winter sun to enter, but prevents overheating from the high summer sun.

glazing. The black paint on the wall enables it to readily absorb solar radiation, and the heavy masonry or concrete construction provides an element of thermal mass to enable the wall to retain heat, which is collected during the day, retained and released throughout the night. As the concrete begins to heat up, some of the heat is transferred into the air between the wall and the glazed façade. This heats up the air, causing it to become less dense and so rises. This sets up convection currents. At the top and the bottom of the wall (and sometimes at the top and bottom of the glazing), there are holes cut into the wall (and sometimes glazing), which are then covered with movable 'flaps', louvres or closable grilles. By adjusting the position (open/closed) of the flap, louvre or grille, it is possible to achieve some degree of control over the heating provided by the Trombe wall.

Night cooling can also be achieved, where the warm air inside the building enters via the top flap and hits the cool glass, causing it to become more dense, which makes it fall and exit via the bottom flap, pulling more warm air in as it does so. Careful use of the flaps prevents this from occurring, where it is undesirable. Additionally, by integrating flaps into the glass façade, it is possible to heat fresh air from outside, or to create forced ventilation by creating a convection current to suck air from the building, heat it and eject it outside – causing fresh air to be sucked into the building through any openings.

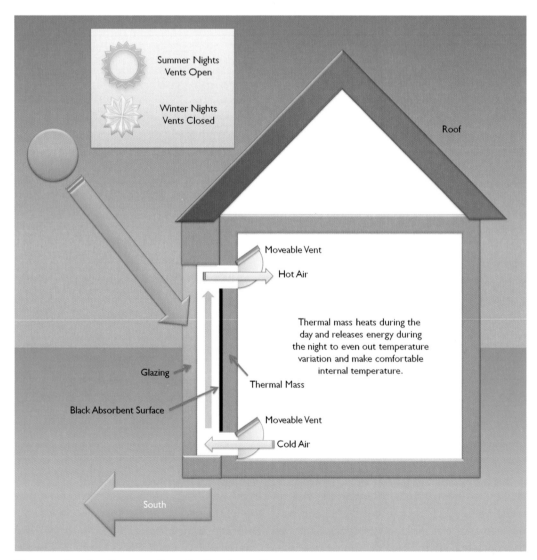

Summer Nights
Vents Open

Winter Nights
Vents Closed

Roof

Moveable Vent

Hot Air

Thermal mass heats during the day and releases energy during the night to even out temperature variation and make comfortable internal temperature.

Glazing

Thermal Mass

Black Absorbent Surface

Moveable Vent

Cold Air

South

Diagram showing the operation of a Trombe wall.

CHAPTER 3

SOLAR POWER FOR LIGHTING

We use an incredible amount of artificial lighting to light dingy corners of rooms in the middle of the day. Principally, this is down to bad building design, which does not make best use of natural light – many problems of having to use artificial lighting during the day can be combated by careful thought at the design stage of how to maximize daylight in the building. We can produce electricity that can be stored or sold and bought back from the grid, to power our lights, but during the daytime, it makes much more sense just to try and get more light to permeate the building – as energy is always lost in conversion. Natural lighting can be used to meet a

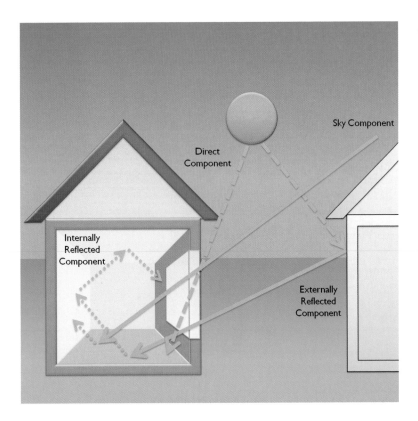

The components of natural daylight.

large proportion of our daytime lighting energy needs and, with effective daylight building design and effective integration with artificial lighting, daylighting can significantly reduce our energy needs.

The Components of Daylight

When considering the daylight that reaches a surface, there are three distinct components to the light, which are expressed in the image on page 31.

The Direct (Sun/Sky) Component

The direct component refers to light that enters the room directly from the Sun or indirectly from the sky. Light that comes directly from the Sun is bright and harsh by nature, while light coming from the sky is diffuse and softer. The 'dome' of the sky makes a vital contribution to the daylight reaching the interior of your property.

The Externally Reflected Component

The externally reflected component refers to all the light from outside the building that is reflected into it. It comprises the light that bounces off adjacent buildings, any walls or other obstructions, trees and foliage, and any light that bounces off the ground, which we refer to as albedo.

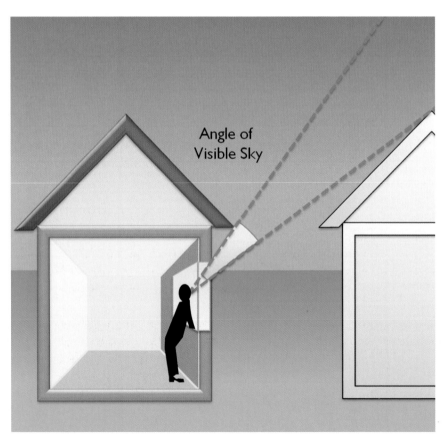

Angle of Visible Sky

The angle of sky visible from this window.

You can maximize the externally reflected component, by ensuring that any surfaces that the daylight must bounce off are painted in bright, reflective colours. If your garden has a large wall, which appears dark and makes the adjacent room look dark, consider painting it white, to maximize the externally reflected component. Similarly, light-coloured flagstones used for paving will reflect more light (albedo) than black tarmac.

The Internally Reflected Component

The internally reflected component refers to all the light that bounces off the internal surfaces of the room before arriving at the work surface.

Angle of Visible Sky

Thorough solar analysis should be conducted, to assess what angle of sky can be 'seen' from all your window apertures. This is the area of sky that is visible when you look out of your window, less any obstructions, neighbouring buildings, trees and so on.

In addition to the 'raw light' entering the room, consideration should also be given to the quality of the light. For certain tasks, it is undesirable for occupants to face 'directly' out of a window, as that would result in unnecessary glare; instead, orient the work-surface at 90 degrees to the incoming light.

Also, consider what areas will be in use during the hours that daylight is available, and try and position these so that they get the best access to daylight. In an existing property, you might consider changing the function of rooms around, to ensure that those rooms that you are using during the day benefit the most from the available daylight.

Daylight can only effectively penetrate the first 6 metres of a room – as the distance from the window increases, the amount of natural light available falls off dramatically. In the pre-electric lighting era, building designers knew this; this resulted in buildings with long, thin shapes, optimized for daylight penetration. We can see in the design of early skyscrapers in the USA, how buildings were effectively designed for daylight penetration and so included features such as atria, which would allow daylight to penetrate the interior of a building. Similarly, if it is a new-build house, try and optimize the rooms' exposure to daylight, and remember the 6m rule – try and optimize your floorplan to make the best use of the daylight – perimeter.

Daylighting Methods

View Windows

Adding a window to a room is a traditional, simple way of adding light to a room; however, there are some limitations with this approach. Windows at eye-level can be a source of glare during the day and provide poor light distribution.

High-Mounted Sidelights (Clerestory Windows)

High-mounted sidelights are windows mounted vertically towards the top of a wall. Clerestory windows are an effective way of getting additional light into a room. A south-facing clerestory window will clearly provide more daylight illumination than a north-facing clerestory. If the windows face south, adequate solar shading should be provided, to minimize unwanted gains in the summer. If possible, brightly painted shading will help the reflected component of the light.

High-Mounted Sidelights (Transom Window) with Light Shelf

In order to extend the amount of daylight from a clerestory window or high-mounted sidelight further into a room, it is possible to incorporate an architectural feature known as a 'light shelf'. A light shelf is a horizontal protrusion beneath the window, which is painted on its top surface with white or another colour/finish with high reflectance. The light shelf should reflect a component of the light back onto the ceiling of the room, providing illumination 'deeper' into the room than could be obtained with a standard window.

Wall Wash Toplighting

Mounting a rooflight adjacent to a wall can provide attractive use of daylight, using the wall to diffuse and scatter the light throughout the room. If this strategy is to be adopted, consider the finish of the wall and optimizing its reflectance for optimal reflection of daylight throughout the room.

Central Rooflights (Toplighting)

Horizontally mounted rooflights can be problematic, due to the fact that they receive most of their light during the period when the sun is directly overhead – noon and the middle of the day – whilst performing poorly at the beginning and end of the day.

The performance of a centrally mounted rooflight can be enhanced by the addition of an optical element – such as mirrors or prismatic devices, which help to capture some of the additional light at the beginning and end of the day that would normally enter the rooflight obliquely and thus not be effectively routed into the room area below.

Patterned/Linear Toplighting

For areas where there is a large space, which requires even illumination, it is possible to fit rooflights at periodic intervals to achieve even illumination over a wide area – where light from a single point source would provide uneven illumination.

Sunpipe

For areas in the centre of a property, where it would be inappropriate to place a window (or inaccessible), a 'sunpipe' is a good way to route natural light into the building. One of the advantages of using a sunpipe – over say, installing a rooflight or an additional window – is that no major structural alteration is required; whereas the addition of a window may require reinforcement of the building fabric or addition of a lintel, the collector of a sunpipe simply fits between the building's existing rafters.

The shape of the collector means that the elements will keep it clean and no additional maintenance or cleaning is required. Contrast this with a horizontal rooflight, which may become dirty and require cleaning in order to maintain an attractive visual appearance and adequate transparency.

For buildings that are in areas that are sensitive to visual appearance, where the dome of a sunpipe might be viewed by planners as too obtrusive, and not in keeping with the character of the area, Monodraught manufacture a 'conservation sunpipe' collector, which replicates the appearance of a cast-iron Victorian rooflight, whilst still providing the flexibility that comes with the sunpipe system.

There are a number of components to the sunpipe system: the externally mounted collector, which harnesses the available daylight, and an ultra-reflective tube, which routes the light

Sunpipe mounting detail. (Monodraught)

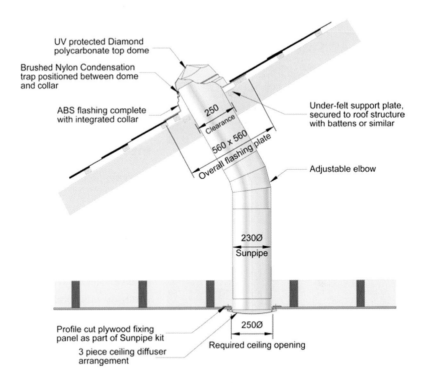

UV protected Diamond polycarbonate top dome

Brushed Nylon Condensation trap positioned between dome and collar

ABS flashing complete with integrated collar

250 Clearance

560 x 560 Overall flashing plate

Under-felt support plate, secured to roof structure with battens or similar

Adjustable elbow

230Ø Sunpipe

Profile cut plywood fixing panel as part of Sunpipe kit

3 piece ceiling diffuser arrangement

250Ø Required ceiling opening

from the collector to the internal luminaire. The pipe is rooted through the cavity in the roof space, so allowing daylight to be 'moved' effectively from one location to another.

The collector comes with a square flashing plate, which integrates with roof tiles; these are supplied in either ABS, galvanized finish or, where a traditional appearance is important, lead.

It is also possible to mount sunpipes in roof areas that might be occupied or serve a mixed use. The photo on page 36 shows how a sunpipe collector can be mounted alongside a rooftop garden, with minimal visual intrusion. For an application like this, where it would be possible to gain access to the roof collector and where security could become an issue, an optional set of stainless steel bars are available, which fit behind the collector, preventing access in the event of a person trying to gain access by removing the sunpipe dome.

There are a number of additions to a basic sunpipe system that provide flexibility for a range of applications with specific requirements. Where fire protection is necessitated, an intumescent collar is available, which fits around the sunpipe above the diffuser. In the event of a fire, this collar will expand, crushing the pipe and preventing the spread of flames and smoke. Furthermore, where privacy is required, acoustic baffles are available, which prevent the spread of sound through the sunpipe system.

Inside the property is a round circular diffuser, which takes the light from the collector and distributes it evenly through the room. Diffusers are also available that integrate an electric light into the fixing – providing supplementary artificial light when daylight is not available.

Sunpipe collectors and diffuser shown in a variety of applications, with the standard collector mounting as supplied. (Monodraught)

Whilst there is limited variation in the diffuser shape and style, a variety of trims in white, chrome or gold are available, to ensure that the diffuser looks appropriate in a variety of décor styles. The illustrations on the left show the sun-pipe diffuser, in a couple of domestic settings.

Interior Glazing

In some circumstances (where privacy is not an issue), daylight can penetrate further into a building by using glazed or transparent divisions, rather than opaque walls. Where some degree of screening is required, frosted glass blocks used as a partition strike a compromise between allowing daylight to reach further into your house, and providing a measure of privacy. Interior windows can also be used to partition spaces physically and acoustically, whilst still allowing light to pass inside the building.

Daylighting Controls

Automatic controls for lighting are commonly used in commercial buildings, but so far have found fewer applications in a domestic setting. However, with increasing focus on energy consumption in buildings, there is a high likelihood that there will be greater scrutiny of domestic-lighting energy use, and homeowners will begin to appreciate the energy-saving benefits that better lighting controls can bring. It is possible to fit a photoelectric sensor, which will sense the amount of daylight that is available and adjust the proportion of light supplied by supplementary artificial lighting automatically. Control can either be on/off or continuously variable. This can be used in conjunction with passive infra-red occupancy sensing, to ensure that the energy use as a result of lighting is kept to a minimum.

If the sunpipe is being used in an area where sometimes it would be desirable to have a lower level of illumination, or no light at all, there are a number of options. Black-out covers are available to shade the diffuser or, for an automated system, a motorized damper is available, which can be 'inserted' into the pipe between the collector and diffuser. Within this damper is a series of baffles that are moved by an electric motor, providing automatic dimming of the natural light illumination level.

CHAPTER 4

SOLAR POWER FOR ELECTRICITY

Solar Power for Domestic Electricity

By fitting photovoltaic cells to your property, your conscience can be as clean as your energy supply. Generating your own solar electricity brings with it the prospect of getting a cheque from your energy supplier, and safeguarding your wallet against future rising energy costs – bundled with the fact that the power generated is carbon-free and not damaging to the environment – what could be more rewarding? This chapter explores producing domestic electricity using solar power.

Domestic Energy Consumption

The domestic sector accounted for 29 per cent of the UK's electricity consumption in 2005. When you consider that you, as a homeowner, account for a fraction of this 29 per cent, and that you can make steps to reduce this 29 per cent by energy efficiency and generating your own, free, solar electricity, then you begin to get a sense of empowerment – you can be part of the solution, not the problem.

Not only can you, as a homeowner, help to reduce the burden of domestic electricity supply by providing some of it from clean, renewable

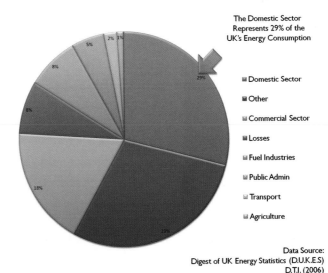

The Domestic Sector Represents 29% of the UK's Energy Consumption

UK domestic electricity consumption by sector.

- Domestic Sector
- Other
- Commercial Sector
- Losses
- Fuel Industries
- Public Admin
- Transport
- Agriculture

Data Source:
Digest of UK Energy Statistics (D.U.K.E.S)
D.T.I. (2006)

energy, but you can also help to reduce the losses inherent in the current centralized generation system (which accounted for 8 per cent of total electricity produced in 2005).

One of the benefits of de-centralized energy production – generating it on-site using renewables – is that less electricity is 'lost' to resistance. By generating electricity near the point of use, rather than far away, we minimize transmission losses. Contrast this with the situation where power is being produced centrally and transmitted a long way by cable – some of the energy is turned into heat as a function of the resistance of the cables and ends up being wasted rather than useful, delivered energy.

Let's take a closer look at how electricity is currently supplied in the UK.

The UK's Electricity Supply

Currently the UK's electricity supply is provided by means of a blend of different technologies – coal, oil, gas and nuclear technologies are all employed, along with a growing proportion of renewable energy technologies.

You will see from the piechart (below) that the UK's electricity predominantly comes from fossil fuels. Nuclear accounts for under a fifth of the UK's electricity supply. However, all this could change in the future as the prices of fossil fuels begin to rise. Predominantly, the UK relies on 'natural gas' to produce its electricity. Cheap supplies of 'North Sea gas' prompted the 'dash for gas', where much old coal-fired plant was abandoned in favour of small, very responsive gas-power stations.

Switching to gas from, predominantly, coal helped the carbon emissions of the UK significantly, as a gas-power station produces less carbon per unit energy than the equivalent coal-fired power station. In recent years, however, the cheap supply of natural gas from the North Sea has begun to run out, and the UK faces fresh new challenges with how to provide its energy. Tough decisions need to be made.

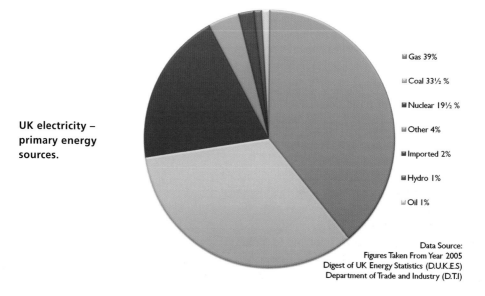

UK electricity –
primary energy
sources.

Gas 39%

Coal 33½ %

Nuclear 19½ %

Other 4%

Imported 2%

Hydro 1%

Oil 1%

Data Source:
Figures Taken From Year 2005
Digest of UK Energy Statistics (D.U.K.E.S)
Department of Trade and Industry (D.T.I)

Unfortunately, when looking at the relative proportion of energy generated by each technology, fossil-fuel technologies (coal, oil and gas) still dominate, with the associated environmental impact that they bring with their associated carbon emissions. Nuclear power, whilst only accounting for a relatively small percentage of UK power production, brings with it a legacy of nuclear waste, which the world still has no idea of how to deal with in the long term. In addition, there are high-voltage cables underneath the Channel, that allow us to import and export electricity from the Continent. Most power imported into the UK comes from the French fleet of nuclear power stations.

Out of all of the technologies, renewable energy technologies have the potential to deliver clean green energy, without the associated environmental costs of fossil fuels and nuclear.

The Photovoltaic Effect and Photovoltaic Materials

It's worth having a basic understanding of the physics behind how the cells on your roof generate energy, because this will help you to understand the characteristics and operating parameters of photovoltaic (PV) cells. PV cells are comprised of semiconducting material. They can be divided into two classes: crystalline and thin-film. We are going to look at the operation of crystalline solar cells in basic detail; looking at what goes on at an atomic level, then looking at their construction and basic properties. Crystalline solar cells are a wafer of solid semi-conducting material, usually silicon.

We can see in the figure above a representation of silicon. In its natural form, silicon forms a lattice structure, which is represented here.

In order to make the silicon 'useful', we 'dope' it with some additional chemicals – this creates N-type and P-type silicon. For N-type silicon, we

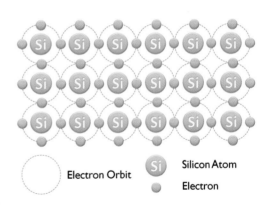

Crystalline structure of undoped silicon.

usually add a little bit of phosphorus to the silicon mix, which adds an extra electron wherever the phosphorus appears in the crystalline structure. Conversely, we can also 'dope' silicon to make P-type silicon, using a little bit of boron. P-type silicon is the opposite to N-type silicon, in that it has a few missing electrons, which we call 'electron holes'. Representations of N- and P-type silicon are shown below and on p.40.

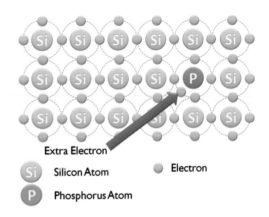

Crystalline structure of doped N-type silicon.

39

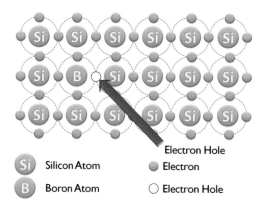

Si — Silicon Atom

B — Boron Atom

● Electron

○ Electron Hole

Electron Hole

Crystalline structure of doped P-type silicon.

In a photovoltaic cell, we create a 'junction' between P- and N-type silicon – imagine a sandwich made from bread, with white bread on one side and brown on the other. This is our solar cell.

Now, light energy from the Sun can be visualized as a stream of photons. When the photons hit the junction, they cause the 'spare' electrons to get excited, raising them to a higher

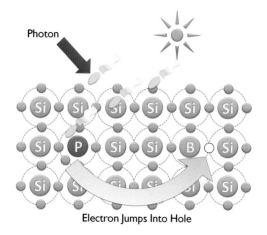

Photon

Electron Jumps Into Hole

Photons hitting the junction cause spare electrons to 'jump' into adjacent holes.

energy level. The electrons then 'jump' across the junction into the electron holes as depicted below.

Crystalline solar cells take this sandwich and add electrodes to either side, allowing us to harness this photovoltaic effect to generate electricity. The electrons flowing around the circuit are the electrons that have jumped across the junction. A region called the 'space charge region' is like the filling in the sandwich. This is the no-man's-land between P- and N-type silicon, where the transfer of electrons takes place.

We can visualize solar cells in an electrical schematic using the symbol shown in the figure opposite. We are going to take a look at how solar cells are connected together to form modules a little later in this chapter, but for now, be aware that the right-hand symbol is sometimes used to denote a solar module.

The Electrical Properties of Solar Cells

To understand a little bit about the performance of solar cells, we need to look at their electrical properties. As we vary the load on a solar cell, we change the current and voltage flow, and can plot this line on a current–voltage graph. The power produced by the cell is the product of the current and voltage.

When we examine the relationship between the current flow and the voltage output of a solar cell, we find that there is an optimum point called the 'maximum power point' at which the load on the cell extracts the maximum amount of power from the cell.

You will notice in the graph on page 42 that there is a square drawn within the graph between the 0 point and the maximum power point – the area of this square is called the 'fill factor'; the greater the area under the current–voltage curve that the fill factor fills, the

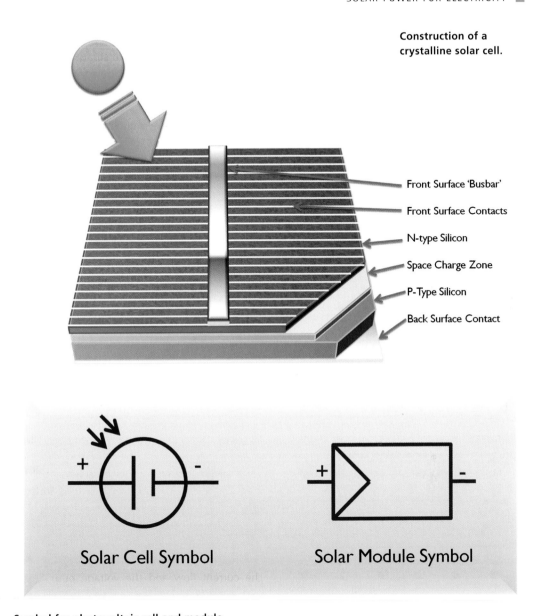

Construction of a crystalline solar cell.

Front Surface 'Busbar'

Front Surface Contacts

N-type Silicon

Space Charge Zone

P-Type Silicon

Back Surface Contact

Solar Cell Symbol

Solar Module Symbol

Symbol for photovoltaic cell and module.

more ideal the performance of the cell. A good quality inverter will 'follow' the maximum power point, to extract the most power from the cell under a range of conditions.

We connect solar cells together in order to generate higher voltages and currents. Where solar cells are connected in series, the voltage output of the string of cells increases; when cells are connected in parallel, the current increases. We can use a combination of cells in series and parallel to generate larger voltages and currents to supply to our inverter. In practice, the situa-

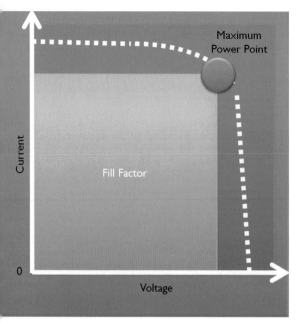

Current–voltage curve showing cell power output and maximum power point.

tion is a little more complex that this – we also include bypass diodes in the strings; this means that, if a cell is overshadowed, rather than becoming a load, the current can flow through the bypass.

We use different terminology to refer to the different components of solar systems. Often these terms are used loosely and interchangeably; however, formally they have different meanings. A photovoltaic or solar cell is used to denote the smallest component of a solar system, a single sheet of silicon or thin film. These cells are in turn connected together to form modules. A module is a unit that provides mechanical support for a number of cells and provides interconnection between cells. A number of modules can thus be combined to form a larger 'solar panel', and panels in turn can be connected together to form a solar 'array', which is the sum total of all the cells used in a single installation.

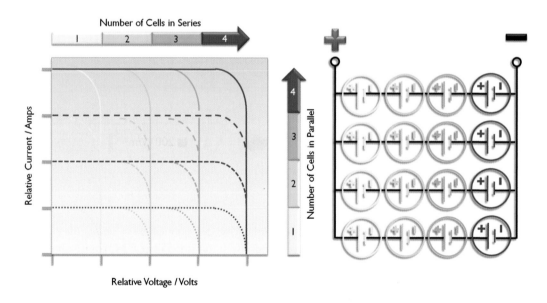

A combination of series and parallel circuits is used to connect cells to form modules.

RIGHT **The relationship between cells, modules, panels and arrays.**

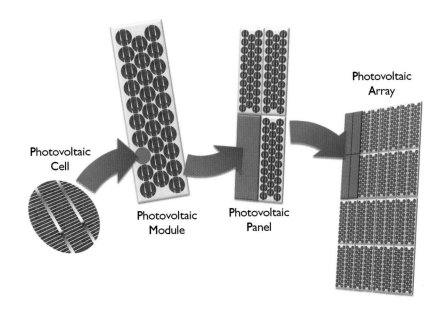

Photovoltaic Cell

Photovoltaic Module

Photovoltaic Panel

Photovoltaic Array

BELOW **How module power increases with increasing irradiance.**

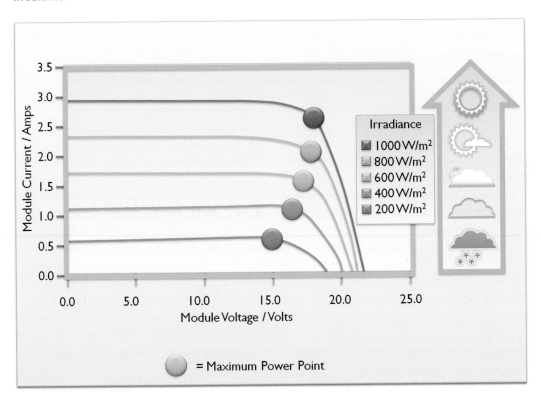

Irradiance
- 1000 W/m²
- 800 W/m²
- 600 W/m²
- 400 W/m²
- 200 W/m²

Module Current / Amps

Module Voltage / Volts

○ = Maximum Power Point

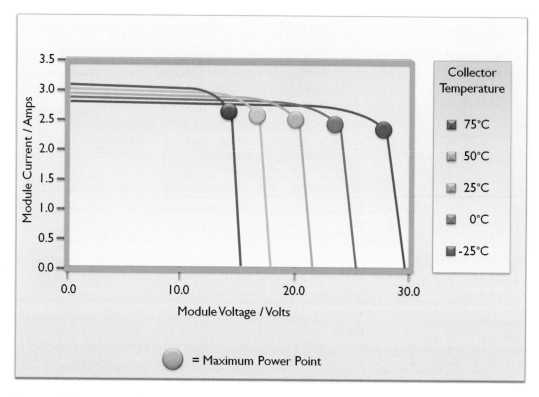

How module power is affected by temperature.

It should become apparent, that as the Sun gets brighter, the amount of power generated by a solar panel increases. The relationship between the current and voltage and irradiance can be seen in the graph on page 43.

What is less obvious is how temperature affects the performance of solar cells. Whilst we are not going to delve into the physics of why this is the case, as a rule of thumb, we can remember that, as the temperature of the panel decreases, the amount of power produced increases (*see* above).

Deregulation is Good News for You

One of the things to have changed significantly about the UK energy supply industry in the past couple of decades is that, beginning in 1990, the electricity industry has undergone a massive restructuring as a result of privatization. The management of the way that the power is produced and distributed to your home has changed radically over this period, with competition being introduced to the marketplace. This 'liberalization' of the energy markets is good news for you as an ethical electricity consumer, as it brings with it a number of benefits:

■ It allows you to exercise consumer choice by buying electricity from 'green' electricity providers – in a competitive marketplace, consumer choice is a powerful tool to influence the companies that supply electricity.

■ It allows you to act as a small-scale 'energy producer' by selling excess power back to the national grid.

This means that, rather than simply being a passive consumer of energy, you can be active in the way it is produced and levy your consumer spending power to make positive change.

Green Electricity Tariffs

Electricity suppliers are legally bound to produce a proportion of green electricity for every unit of electricity they sell to consumers under the renewables obligation. This is a relatively small fraction when you consider the massive damage that is generated by the other fraction of the unit from dirty electricity sources. In 2006–07 the target for power to be generated from renewables was 6.7 per cent; this is due to rise incrementally every year, with a target of 15.4 per cent for 2015–16.

You may currently be deciding whether to install solar photovoltaics on your property or not; however, without making the commitment of finance and energy in altering your property, you can make the commitment towards using 'green energy'.

In addition, it is worth considering that, if you are planning a grid-connected solar electricity system, the system may only provide part of

If you are interested in finding out how much the switch to green electricity will cost you, check out www.greenelectricity.org, which allows you to input your annual or quarterly energy usage along with your postcode, to calculate how much your supply would cost with different suppliers.

your energy needs and there will be times when there is insufficient solar energy available to meet your electricity requirements – then it makes good sense to have a 'green supplier' to meet the shortfall.

Liberalization of the UK's energy market has resulted in increased choice for consumers as to who supplies their energy. No longer are you bound to a 'local' supplier. There are a number of energy suppliers who will guarantee that, for every unit of electricity you consume, the supplier will source this electricity from a renewable energy supplier. By buying 'green' electricity you are:

■ Demonstrating that there is a clear consumer need for ethically produced power – this sends a clear signal to companies producing electricity.
■ Providing funding for clean-energy projects and more renewable infrastructure to be built.

Will it Cost Much More?

Some green energy suppliers will charge you no more than their 'standard' rates for electricity – with the 'catch' being that you may not be able to take advantage of some of their 'special offers' or 'cut-price' tariffs. Other suppliers may charge

THE GREEN ENERGY MIX

It is worth bearing in mind that, if you choose a green energy supplier, the energy is most likely to be generated from wind farms and hydro-power schemes, as these are presently the easiest way to generate large-scale renewable energy.

GREEN ELECTRICITY SUPPLIERS

Ecotricity: www.ecotricity.co.uk

Good Energy: www.good-energy.co.uk

nPower Juice:
www.npower.com/at_home/juice-clean_and_green.html

Powergen GreenPlan:
www.powergen.co.uk/Products/Products-Electricity-And-Gas/Products-GreenPlan.htm

Green Energy UK:
www.greenenergy.uk.com

a small premium for 'green energy'; if, for example, you pay between 7 and 10 pence for each 'standard' electricity unit, you might expect to pay between 8 and 18 pence for each 'green' electricity unit. Additionally, some suppliers offer a scheme where you purchase green electricity on a 'tiered' system, where the first X amount of units is charged at a higher tariff, and the remainder at a lower tariff.

What is the Difference Between a 'Green Supply' and a 'Green Fund'?

If you receive your electricity from a 'green supplier', then your supplier pledges to match each unit of electricity that you purchase with one provided from a renewable source. With 'green funds', your supplier pledges that a proportion of the money you pay for your bill will go towards green-energy projects.

Assessing Your Energy Needs

Before you begin to look at meeting the needs of your house from renewables, you need to assess carefully your total daily energy needs. The following table should prove useful in helping you to do that.

Power Saving and Reduction

The first step in moving towards making your home energy supply 'greener' is to look at your energy use. At the moment, home energy use is profligate, with so much unnecessary consumption, which is a result of the bad 'energy habits' that we have developed.

Carry out a personal energy audit to see areas where you can save energy. Some ideas are listed below:

■ Be ruthless with any old incandescent lightbulbs. Compact fluorescent or 'energy-saving' lightbulbs are available in a wide variety of styles and have shaken off the image of taking a long time to 'warm up' and of being 'flickery', as a result of modern electronic starters. A relative newcomer to the market is LED lighting, which offers a plug-in replacement for feature lighting such as 'halogen' lights commonly recessed in the ceiling or used for feature lighting.

■ Next, take a look at all of your old appliances. Think carefully about how much energy they are using. If you are concerned, you can buy a plug-in meter, which goes between your appliance and the socket. When replacing appliances, look for those that are energy-efficiency 'A' rated; over their life-time, they should pay for any additional money spent in purchase in avoided energy costs. When buying washing-machines or dishwashers, look for those with an 'economy' or 'light' setting.

■ Also, did you know that every time you open an oven door, you lose 25 per cent of

Calculating Total Daily Power Consumption

Device	Load (in Watts)		How long do you use it for?		How many days of the week?		Watt-hours per Day
	W	x	:	x		÷7=	Wh
	W	x	:	x		÷7=	Wh
	W	x	:	x		÷7=	Wh
	W	x	:	x		÷7=	Wh
	W	x	:	x		÷7=	Wh
	W	x	:	x		÷7=	Wh
	W	x	:	x		÷7=	Wh
	W	x	:	x		÷7=	Wh
	W	x	:	x		÷7=	Wh
	W	x	:	x		÷7=	Wh
	W	x	:	x		÷7=	Wh
	W	x	:	x		÷7=	Wh
	W	x	:	x		÷7=	Wh
	W	x	:	x		÷7=	Wh
	W	x	:	x		÷7=	Wh
	W	x	:	x		÷7=	Wh
	W	x	:	x		÷7=	Wh
	W	x	:	x		÷7=	Wh
	W	x	:	x		÷7=	Wh
	W	x	:	x		÷7=	Wh
	W	x	:	x		÷7=	Wh
	W	x	:	x		÷7=	Wh
	W	x	:	x		÷7=	Wh
	W	x	:	x		÷7=	Wh
	W	x	:	x		÷7=	Wh
	W	x	:	x		÷7=	Wh
	W	x	:	x		÷7=	Wh

Total daily power consumption _____

the heat? This heat requires energy to replace. Also, consider leaving the oven closed, turning the heat off before the food is cooked and allowing a little longer. The heat in the oven should be enough to finish the cooking if it is left for a little longer to compensate.

- Let things cool down thoroughly before putting them in the fridge or freezer – as putting something warm in will use additional energy to extract heat. If the cooling coils, hidden behind your refrigerator are dusty or dirty, give them a spring clean – as this grime will impede efficiency. Check the gasket on your fridge and freezer to ensure a good seal. If cold air can escape, replace the gasket to ensure your fridge or freezer works at optimum efficiency. Also, consider turning the temperature inside these items 'up' and placing items which require lower temperatures near the coldest part of the fridge.

- Electric heating is a big no-no. It takes a massive amount of electricity to produce heat – it is much kinder to the environment to heat with wood or gas.

- Power showers will also sap electricity, so you might consider using a water-pressure shower.

- Few properties in the UK have air-conditioning fitted as standard, although the recent spate of hot summers may lead many to believe that we will increasingly see air-conditioning becoming common in the UK. Air conditioning wastes a tremendous amount of energy – it is much more environmentally friendly to open a few windows and install a ceiling fan, if extra cooling is needed.

- Also, make sure that all electrical goods, such as TVs and videos, are switched off 'at the wall' rather than leaving them on standby (which saps electricity).

- Ensure that, if you are going to have lighting outside, it is fitted with PIR sensors (passive infra-red), which detect the presence of people. This will ensure that they aren't inadvertently left on, if you forget to turn them off.

Assessing Your Home's Potential for Solar Electricity Generation

Fitting Solar Cells to Existing Properties

If you are going to make a serious commitment to installing solar electricity generating capacity in your home, the first thing you need to do is conduct a survey of your home and assess its suitability for solar electricity generation. If you have plans of your home or construction drawings, you can glean a lot of useful information from them about:

- the roof area that is available for installation of solar photovoltaic modules;
- useful information about the slope of your roof, which will help you to determine how efficiently your panels will be able to harness solar energy at different times of the year;
- the construction of your roof, which will help you to determine whether your roof can support the additional weight of solar photovoltaic panels.

New-Build Properties

If you are building your house and it is still on the drawing board, you have much more flexibility in considering your options for solar photovoltaic generation; you can take into account the optimum roof angle at an early stage and design this into your property – even if your funds do not permit installation of solar photovoltaic cells at the moment, you can 'design for the future', taking into account the requirements of solar PVs and allowing you to retrofit them at a later date.

Shading Evaluation

Any form of shading, or anything that obstructs your PV modules' access to the light is bad news, as it has the potential to significantly impact on the yield of power from your PV array. Considering that your PV array is a significant investment, it is sensible to invest the time to ensure that there is nothing that will unduly impede the performance of your system.

Permanent Shading

Permanent shading is caused by objects that are permanently in the way and cause a shadow on your PV modules. You should look for features of the building, such as walls, dormer windows built into the roof structure, chimneys and other buildings, which could cast a shadow on your PV array. The Sun changes its position in the sky throughout the year, as we have seen, so how permanent objects impact the shading of your PV modules will change over the course of a year.

Natural Obstacles

Also, look for natural obstacles such as trees – and remember that these will grow over time, so whilst they might not overshadow your PV array when it is installed, check every year to ensure that any new growth does not encroach on your PV's sunlight.

Temporary Shading

Any detritus that collects on your PV modules and prevents them getting a clear view of the Sun, will affect the performance of your array. By tilting your modules, a 'self-cleaning' effect is achieved, whereby the rainwater washing over them allows them to cleanse themselves of any accumulated dirt and debris. An angle of at least 12 degrees is sufficient to achieve self-cleaning – most PV arrays in the UK will be oriented at a steeper angle for reasons of efficiency anyway, so our arrays should cleanse themselves with relative ease.

Leaves

If there are a lot of deciduous trees in the vicinity of your installation, the leaves that fall off the trees in autumn have the capability to collect on your PV module. A slope on your PV modules should ensure that the wind and rain keep your PV panels clean and that any leaves quickly blow off. However, if you notice an accumulation of leaves that do not clear, you may have to remove them manually.

Snow

Snow has the potential to obscure a PV array's access to the sun; this will be of more concern to those at higher latitudes and altitudes in the UK. Again, the natural pitch of your PVs should ensure that snow clears quickly.

Soot and Dust

If you live in an area that is largely industrial (or if you generate heat or power from burning oil or coal), then soot and dust can collect on your PV array and impede its operation. Think carefully when siting your PVs that they are not situated near the exhaust of any diesel-generator sets or the flue of oil-burning boilers, as these present a ready source of soot and dust, which will degrade the performance of your modules if allowed to accumulate over time.

Comparison of Different Photovoltaic Technology Types

Cell Material	Efficiency	Area Required to Generate 1kW Peak
Monocrystalline Silicon	15–18%	7–9m²
Polycrystalline Silicon	13–16%	8–11m²
Thin Film Copper Indium Diselenide (CIS)	7.5–9.5%	11–13m²
Cadmium Telluride	6–9%	14–18m²
Amorphous Silicon	5–8%	16–20m²

Source Data: Deutsche Gesellschaft fur Sonnenenergie e.V.

Types of Solar Cell

There is a bewildering array of different PV-cell technologies out there, some technologies are more efficient than others – but that efficiency generally comes at a price. There is a trade-off to be made when considering the appropriate cell technology – you will need to consider how much power you need and how much roof area is available to you.

PVs can be broadly grouped into two classes: crystalline cells and thin-layer cells. We will look at some of the technologies classed under these two headings.

Crystalline Photovoltaics

After oxygen, silicon is the next most abundant element on earth. Most of the silicon in the earth is bound up with oxygen in a chemical called silicon dioxide (SiO_2), which we commonly call 'silica'. When you realize that silica is the main constituent of sand, you realize that we have a lot of it! Of course, in order to make PVs we have to process the silica to make silicon of a very high purity in order to make solar cells. This silicon can then be processed in a number of ways to make the following types of cells:

- monocrystalline (single crystal) solar cells;
- polycrystalline solar cells;
- polycrystalline ribbon cells.

Monocrystalline (Single Crystal) Solar Cells

Monocrystalline photovoltaics are so-called, because they are made from a single crystal of silicon, sliced into wafers. The silicon is drawn from a bath of molten silicon, in a process known as the czochralski process. The large crystals are then cut into slices, called wafers, 3mm thick. The wafers are then 'doped' – a process

that creates a semiconductor junction, and contacts are applied to the cell. The process results in cells that vary in colour from dark blue to black. An anti-reflection coating may be applied, which will give the cells a grey hue.

Polycrystalline Solar Cells

Polycrystalline solar cells are so-called because they are comprised of many crystals of silicon. The process differs from the monocrystalline process, as the silicon is 'cast' into blocks, which results in many crystals forming within the slab of silicon as it cools from its molten state. The blocks, or ingots, of silicon are sliced into wafers, again about 3mm thick. As with monocrystalline cells, the wafers are doped and contacts attached. The process results in cells that are blue in colour. Again if an anti-reflection coating is applied, the cells will appear grey.

Polycrystalline Ribbon Cells

One of the disadvantages with producing the above crystalline solar cells, is that when machining and cutting is performed on the raw silicon, to achieve the uniform size of cell from the raw crystal, much of the silicon is lost as 'off-cuts and dust'. One of the ways of eliminating this dust is to 'draw' the silicon as a ribbon, out of a bath of molten silicon. The ribbon is already the thickness required of the finished solar cell – so it just needs to be sliced to size, producing less wastage. Ribbon cells tend to be blue in colour.

Thin-Film Photovoltaics

Thin-film cells differ from crystalline cells in that, rather than being comprised of a solid piece of semiconductor (as in the above examples), thin-film cells consist of a supporting substrate, such as glass or polymer, on which a thin, film coating of semiconductor is deposited. This results in cells that are much thinner, use much less material and require much less energy in manufacture. The process offers the flexibility of being able to make cells of different sizes. Rather than electrical contacts being 'attached' to the cell, the contacts are deposited onto the substrate. Typically, an opaque metal coating is used; however, for the front of the module, a transparent metal oxide conductor is employed, which is deposited on top of the thin film.

Amorphous Silicon

Amorphous silicon solar cells differ from mono- and polycrystalline cells in that, rather than having any defined crystal structure, the silicon forms an irregular network. Amorphous cells are made using a vapour-deposition process. One of the disadvantages of this type of cell is that, within the first year, the performance of the cell rapidly degrades resulting in less power being generated, but after this period, the cell settles to a stable level of power generation. Amorphous silicon cells can range from being a red-brown colour to black.

Copper Indium Diselinide (CIS Cells)

So far we have dealt with solar cells that employ silicon-only as the semiconductor. Copper indium diselinide cells, as should seem apparent from the name, do not use any silicon. Unlike the amorphous silicon cells, CIS cells do not degrade in the first year – they are stable. CIS cells are not particularly suitable for environments where high temperatures or humidity feature. CIS cells are presently the most efficient out of all of the thin-film technologies. CIS cells are black.

Cadmium Telluride

Like CIS cells, cadmium telluride cells do not use silicon as the semiconductor. The semiconductor is deposited onto the substrate by a process called 'vapour deposition'. One of the considerations that should be borne in mind if selecting these cells is that they contain cadmium. The cadmium telluride compound itself is non-toxic and stable; however, during the manufacturing process, cadmium is used that is toxic – particularly during the process of deposition, when it is gaseous. Cadmium telluride cells vary in colour between a dark green and black shade.

Solar-Panel Mounting Options

The decision of where to mount your photovoltaic array will depend largely on the constraints of your site, and what space and sun-facing resource you have available. Many installations use the roof-space of a house or extension to mount photovoltaic panels; however, there are other options available to creatively use solar panels in your home. Before beginning to plan where to site your solar array, ask yourself some of the following points.

Is your building surrounded by an expanse of land with favourable solar properties suitable for mounting a stand-alone solar array?

Photovoltaics can be creatively integrated into vertical façades, providing a way to renovate buildings with ageing exterior finishes. The image opposite shows how solar cells have been used in Germany to renovate this block of flats – the ageing concrete façade has been replaced in these homes with jewel-like solar panels, enhancing the appearance of the building, whilst generating clean energy.

There are all sorts of opportunities for creatively integrating photovoltaic panels in your building; for example, a solar panel may be mounted over a window to provide solar shading at the same time as energy generation.

Why not combine sustainable energy transportation with sustainable energy generation? A bicycle shelter provides an ideal location for mounting solar panels, whilst keeping your cycle dry.

RIGHT **Solar cells mounted vertically on façade.**

BELOW **PVs mounted above window additionally provide solar shading. (Maria Hawton-Mead)**

This bicycle shelter provides both sustainable transportation and sustainable energy. (Dulas Ltd)

Carports, lean-tos and other areas that would normally employ a large area of flat roofing, can also provide opportunities for mounting solar panels.

Solar Trackers

If you have a property with a large amount of open space, which is not overshadowed significantly by trees or buildings, and which is facing the right direction, you might want to consider a solar tracker.

As we saw looking at the Sun-path diagrams earlier in this book, the Sun does not occupy a constant position in the sky – it moves through the sky, rising in the east, and setting in the west – if you are in the Northern Hemisphere. This means that if we mount solar panels in a fixed orientation, we do not harness the maximum amount of energy possible, by virtue of the fact that sunlight hits the panel at an angle through most of the day. This results in our solar panel not harnessing the maximum energy that it could if it were facing the Sun constantly all day, every day. There is a way to overcome this though – a solar tracker. Solar trackers allow the solar panel to move – generally rotating about one axis, rotating through azimuth. However, some allow the panels to track the Sun in two axes of rotation – azimuth and elevation. The additional complexity inherent in such a design can count against it; in many cases, a one-axis tracker is more than sufficient to accomplish the task.

There is a small amount of energy required to drive the motors that allow the panels to track the sky. However, with careful design, the benefits derived through increased energy generation offset the small amount of energy required to drive the tracker circuitry. The benefit of a tracker is felt much more in the winter months, where the improvement in irradiance can be as much as 300 per cent at some times of day. In summer, the gains are still appreciable; however, they fall down to a more modest 50 per cent. Over the course of a day, a single-axis tracker will capture between 15 and 25 per cent more power over the course of the day, with a dual-axes tracker delivering an additional 5 per cent.

A commercially obtained solar tracker will come with integral drive electronics, which will drive the motors that change the orientation of the solar panels. This may require extra wiring for installation, entailing additional costs, and should be enquired about before deciding on a solar tracker system.

An interesting variation on this theme has been executed by a homeowner in Germany, where a rotating roundhouse has been built with photovoltaics integrated into the design. The whole house rotates to track the sun.

Roof-Mounted Solar Cells

The simplest way to add solar cells to an existing

A solar tracker in Llanrwst near Snowdonia. (Dulas Ltd)

A solar tracker in Scotland on the property of Mr Howie. (Dulas Ltd)

BELOW Solar-tracking roundhouse In Germany. (Maria Hawton-Mead)

building is to use mounting hardware and mount solar PV panels to the roof of the building. Many buildings in the UK have a suitable roof pitch to be able to collect usefully solar energy. One of the disadvantages of this method is that retrofitted solar panels do not 'blend in' very well with the roof line, and can appear a little obtrusive; however, this small visual disadvantage is far outweighed by the promise of clean, green energy being generated by the panels.

Roof-Integrated Solar Slates

PV panels can be mounted over the top of existing roofing structures as we have seen. Another

Housing estate in Thurrock where solar panels are retrofitted to existing roof structure. (Dulas Ltd)

method for mounting PV cells is to integrate them into the fabric of the roof structure. This has the advantage that the PV module now serves two purposes: generating electricity and also taking over the role of the roofing material that it replaces – insulating the building from the ingress of water, weather and the elements. Individual modules can then be connected together to form larger panels, which form part of the roofing structure. There are a number of advantages to this method:

■ Less material is used as the PV module takes on the role of the roof 'slate' it replaces. This results in cost-savings and, from an environmental point of view, energy and material savings, as less manufactured product is required.

■ The building is visually more attractive – as the PV modules seamlessly integrate with the roof, rather than appearing 'bolted on'.

■ The panel, which is 1,240mm wide and 420mm tall, can be installed by any competent roofer, using the same skills as for handling traditional tiles. The tiles are available using two different technologies:

- a polycrystalline tile is available which produces 40Wp and is guaranteed to produce 90 per cent of its power output in 10 years time.
- a solar thermal collector.

Installing Solar Slates

There are a number of different ways of installing solar slates on to a roof; different manufacturers have come up with their own proprietary systems for integrating PV modules with their roofing products. One system uses tiles that screw onto the roofing battens in a similar manner to conventional tiles. The tile, supplied by Solar Century, has the same dimensions as four standard roof tiles and weighs 10kg, which means that the module can easily be carried by a single person as shown in the illustration below.

Before finally fixing the tiles to the roofing battens, they can be hooked into place, like conventional tiles, allowing the fit and spacing to be adjusted.

The tiles require no specialist fixings, but stainless screws are highly recommended to prevent corrosion and damage to the tiles over time. As a result of this, no specialist tooling is required, which keeps costs down. Notice that the cells only cover the 'bottom' area of the tiles and that the top area, for overlapping tiles, contains no cells.

As with conventional tiling systems, you start first at the bottom of the roof and work your way

Installing C21 solar slates step 1.
(Solar Century)

Installing C21 solar slates step 2.
(Solar Century)

up the roof, overlapping tiles as you go. As pointed out, the area that overlaps contains no PV cells, so the top half of each tile can be covered by the tile above.

The solar slates are designed to sit flush with the roofline created by standard tiles, and as a result, no specialist flashing is required to integrate the tiles into the roof. Once installed, the tiles are as weatherproof as any other tiling system. The installation must be completed by a qualified electrician. The panels come with a modular connector system, which means that they can easily be connected together inside the roofing cavity.

The images on this and the next page show a residential development with C21 solar slates – it can be seen how this is a solution that is visually unobtrusive, and can barely been seen from the exterior.

TOP RIGHT **Installing C21 Solar slates step 3. (Solar Century)**

BOTTOM RIGHT **Installing C21 solar slates step 4. (Solar Century)**

59

LEFT **Installing C21 solar slates step 5. (Solar Century)**

BELOW **A residential development with C21 solar slates. (Solar Century)**

A LaFarge roof-integrated PV installation on a new build in Castlemead. (Dulas Ltd)

The LaFarge installation (above) shows a different type of roof-integrated photovoltaic panel installation. In this case, the dimensions of the solar panels are larger than the size of the roof tiles they are designed to integrate with, which results in a different, slightly more defined appearance to the installation – although the benefit of large tiles is that they require fewer interconnections and may reduce labour costs due to quicker installation.

Components of a Solar System

The photovoltaic cells on your roof are just one component in an electrical installation that provides electricity to suit the appliances in your home. In this section, we are going to look at some of the additional components you require, to make a complete solar installation.

There is a variety of ways that a solar system can be configured, which will be discussed in the next section. However, for now we will be examining the components of the system, so that you are familiar with their function when we begin to discuss them.

> For the technical detail on executing a safe solar installation, reference should be made to the IET/IEE Wiring Regulations (17th edition or most current edition at time of reading) BS7671:2008, which now include revised guidance in Part 7 for correct electrical installation of solar equipment.

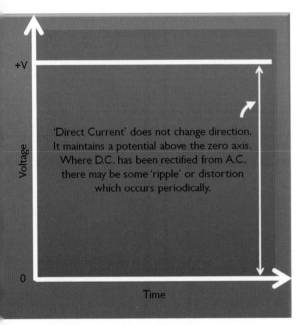

Voltage–time graph showing direct current.

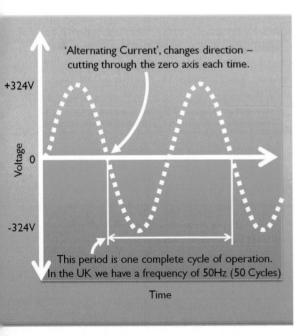

Voltage–time graph showing alternating current.

Inverter

What is an Inverter?

The inverter is the crucial piece of hardware that takes the power provided from the solar cell array, and changes and conditions it to be suitable for domestic mains supply, which it does by using an electronic switching process.

What Does an Inverter Do?

Solar photovoltaic cells produce direct current at low voltage. This is like a battery that produces a low-voltage supply. Direct current (DC) means that the current flows in one direction. The voltage is constant and does not change. We can represent steady DC on a graph by drawing a straight horizontal line as shown (left).

Typically, solar cells are connected to provide a relatively low DC voltage at a high current.

When you plug an appliance into a UK wall socket, the supply that comes off the wall is 'alternating current' (AC). Rather than being constant and flowing in one direction, like direct current mentioned above, alternating current is a constantly varying voltage – however, the voltage varies to a regular pattern at a regular frequency. The pattern that the voltage follows is known as a 'sine wave', as illustrated in the graph (left).

In the UK, the wave 'fluctuates' at a frequency of fifty times a second. The amplitude – that is to say the 'height' of the peaks of the sine waves – corresponds to the voltage of the supply. Following EU harmonization, the supply of the voltage in the UK and the rest of the EU has been normalized to 230V.

With this knowledge, you could easily believe that the 'peak voltage' – or the 'top' of the sine-wave curve – was at 230V. In practice, it is not as simple as this. The figure of 230V is what is called the 'RMS' or 'root mean squared' value of

> For a sinusoidal wave form, the 'RMS' value is equal to the peak voltage multiplied by 0.7. The 'peak voltage' value is equal to the 'RMS' value multiplied by 1.4.

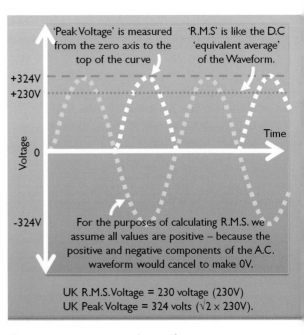

The root mean square voltage of an AC waveform.

the voltage. The amount of 'delivered' power is comparable to that of a supply of 230V DC supply – so the 'RMS' value allows us to make comparisons between AC and DC; however, the 'peak' of a 230V mains supply is around 320V. This is illustrated in the diagram (right), which shows the level of RMS voltage compared to the peak of the sine wave.

Now we understand the difference between the supply provided by the solar cell array and the mains supply, we can begin to appreciate the difficult task that the inverter has of changing one type of supply to another.

Now we are going to look at some more inverter terminology, which is crucial to get to grips with the technology.

True and Modified Sine-Wave Inverters

How accurately an inverter manages to replicate the true shape of a sine wave is a mark of its sophistication and quality. For applications where electric motors are not used, a little quality of supply can be sacrificed in favour of simpler electronics. Rather than 'true' sine waves, some cheaper inverters produce 'modified' sine waves, as shown in the diagram (right). This can be thought of as a 'squarer chunkier' version of the 'true' sine wave. One of the disadvantages of modified sine-wave inverters is that they are not as efficient as 'true' sine-wave inverters. Some modified sine-wave inverters also generate an audible 'buzz' in operation, which can be irritating.

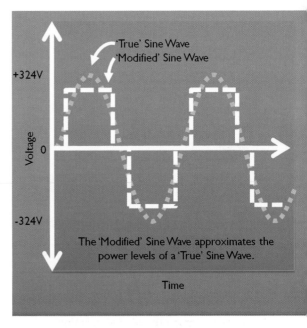

Contrasting true and modified sine waves.

For applications using electric motors, it is best that a 'true' sine-wave inverter is used. They produce a more accurate replica of the sine-wave waveform – this allows motors to run more smoothly. The disadvantage is that true sine-wave inverters are more complex, resulting in greater cost; however, they are silent in operation and more efficient.

Stand Alone, Synchronous or Multifunction

Stand-alone inverters take DC low-voltage power and turn it into AC high-voltage power. However, stand-alone inverters can only work in isolation – they cannot synchronize their supply with another AC supply from the electricity supplier. For this application, a synchronous inverter is required. A synchronous inverter 'looks' at the waveform coming from the incoming electricity supply and matches its output waveform with that coming from the electricity supplier. There is another type of inverter that permits the benefits of both types of system – the multifunction inverter. The multifunction inverter provides all of the functions of synchronizing the AC output with the grid when power from the grid is available, but allows the property to maintain autonomy by permitting a 'backup' bank of batteries and/or generators.

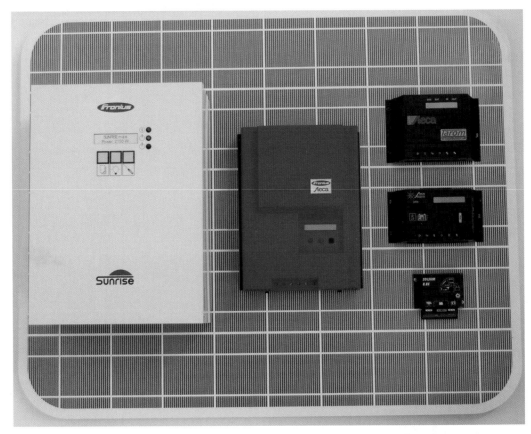

A range of inverters.

ABOVE Weatherproof inverter mounted externally. (Maria Hawton-Mead)

BELOW A typical inverter setup.

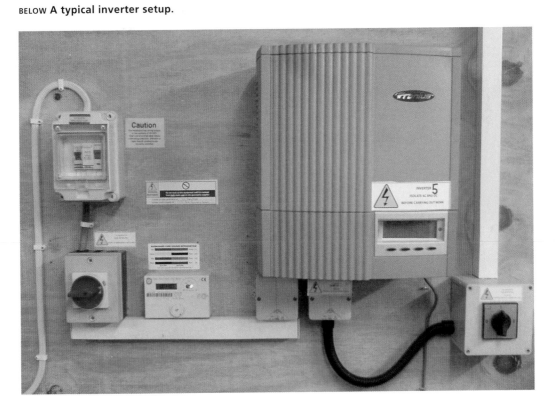

Isolator switch.

Whilst some inverters require mounting indoors in a waterproof location, it is also possible to source inverters that are suitable for mounting externally. This can keep the DC connections between the solar panels and the inverter short – aiding resistive losses – and can be useful in some applications where finding a suitable place inside would be difficult.

Where an inverter is mounted internally, it is usual to include all the associated equipment in a convenient central location. Isolator switches and metering can all be included in a common location, providing easy and convenient access for maintenance.

For a safe installation, isolation and trip devices will need to be provided. The IEE/IET Wiring Regulations (17th edition) contain guid-

Watt-hour meter.

Inverter Size Calculation		
	Line total	Subtotal
Add the total power rating of all the appliances, lighting and other electrical loads you intend to use simultaneously.	W	W
Add the surge power rating of any electrical motors, which may be started (e.g. lifts, hoists, washing machines).	W	W
	Total	W
Now multiply the total from above by a factor of 1.2 to take into account inverter losses.	Total with inverter losses	W
The above is the minimum size of inverter that you will need. However, if you anticipated future power demand, you may discover that you need a larger inverter in the future. You might like to consider adding to the figure above to take into account future-proofing your system.		

ance on the correct way to execute a solar installation, and advise on the appropriate selection of protective devices. A qualified solar installer or electrician will be familiar with this information and should be able to advise.

Isolation should be provided on both the AC and DC sides of the inverter to allow for safe isolation of the circuits for periodic maintenance and servicing. Isolator switches should be fitted with a 'lock off' that allows them to be padlocked off in a 'safe' position, to ensure that they are not inadvertently turned on by an unauthorized party during periodic maintenance.

Means should be provided of monitoring the amount of energy generated by the installation. Metering should be provided by the utility company if the installation interacts with the grid. Additionally, a Watt-hour meter can be installed for the user's own personal information.

Sizing Your Inverter

Now that you have an understanding of what sort of inverter you will need, you can begin to consider what power-rated inverter you will require. To obtain an estimate as to what size inverter you will need, use the table above.

Occupant Feedback Meters

Metering for your electricity supply is essential, so that your supplier knows how much power you are using or providing. However, metering is often shut away, considered an eyesore, hidden under the stairs or with a cupboard built around it to shut it away – apart from that biannual event when the chap comes round to read the meter.

Energy is so important, why do we choose to ignore and pay scant attention to how we use it? Most of us are barely cogniscent of how much energy we use. When you consider that most people pay their bills with direct debit – many of us barely know how much energy we are using and how much it is costing us. If you are going to the expense of installing a solar array, you might like to remind yourself how well you are actually doing, and the positive impact that you are having on the environment.

It is possible to install a meter that is designed to be 'public facing', rather than tucked away in the corner. Such meters might tell you any number of statistics, from the power generated at that moment, to the power generated over a given period of time – or since the system was installed. Some meters even give an approximation of the amount of 'carbon emissions' saved as a result of the installation – over using conventional power from fossil fuels. Such signage is generally used as a tool in schools or public buildings, where they are used to educate the public; but there is nothing to stop you having one to show the neighbours! Anything that promotes solar power and clean energy is to be seen as a move in the right direction.

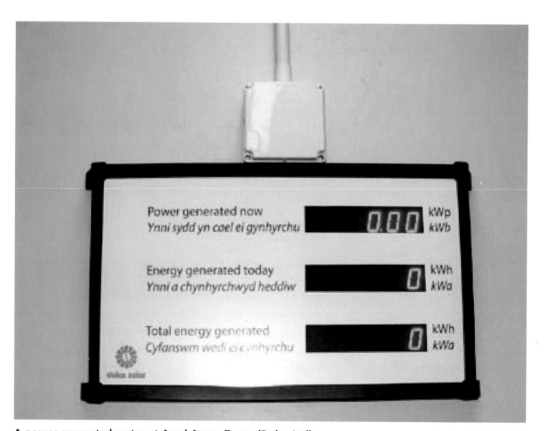

A power generated meter at Awel Aman Tawe. (Dulas Ltd)

The simplest solar setup.

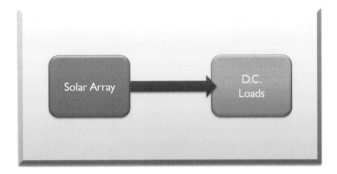

Equipment Configurations

There is a variety of ways that you can configure your solar setup – each with its own advantages and disadvantages. Setups vary enormously in complexity and cost – so you will need to give careful consideration to the resources available when planning your setup.

On-Grid or Off-Grid

The main distinction that we can make between solar setups is whether they are connected to the National Grid or not. The primary factor that will govern this decision is whether a connection to the grid is available locally or not. If the solar installation is for a current domestic dwelling that already has an electricity supply, then a grid-connection is by far the most sensible option. Instances where an off-grid system might be suitable is if you are trying to power a dwelling in a remote rural area where an electricity supply is unavailable, or if you are making a small-scale system to provide lighting and possibly power for a few small DC loads, for a shed or summerhouse (where the cost of a small lead acid battery will be far less than running cabling down the garden). Many people harbour an idyllic dream of living 'off the grid', not dependant on or connected to any kind of infrastructure; however, there are many advantages to

being connected to the grid if the connection is available.

We are going to look at a variety of different configurations, which will suit applications that you may encounter in a domestic setting, from the simplest possible setup to more advanced setups, which interact with the National Grid. First we will look at a few DC-only, off-grid setups, which will be suitable for small applications – such as powering a light and a few small DC loads in your shed or summerhouse. Then we will look at bigger setups, which are suitable for the off-grid home. Then we will look at grid-connected setups.

Solar Installations for the Off-Grid Home

Starting with the simplest and working our way up to the more complex, we will look at installations for the off-grid home. The simplest possible solar setup is illustrated above.

In this instance, a solar cell is connected to a DC load directly with no form of regulation or storage. Think of it a bit like a calculator – you can only use the device when there is sufficient solar resource available. The power quality to the DC load will be low – there is no form of regulation or protection. Such a setup might be suitable for a small-scale water-pumping installation – say for a pond pump or garden irrigation, where the intermittency of the supply will not

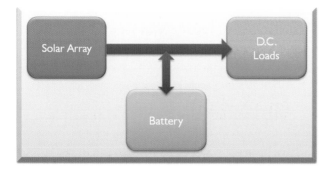

LEFT **Basic off-grid setup with battery storage.**

BELOW **Basic DC off-grid setup with regulation.**

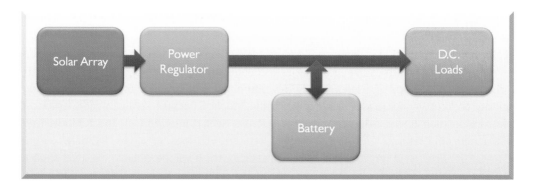

prove a problem and the speed that the water is pumped is not too critical.

We can improve upon this setup slightly by adding a battery to the setup in order to store excess power when it is not being used and to provide power when it is available; this is illustrated above (top).

This sort of installation would work for something like a garden shed or summerhouse, where a simple load such as a light-bulb is being powered. Whilst this solution is elegantly simple, it has a number of caveats. No regulation of the output is provided from the solar panel – as a result, the batteries may not be charged in the optimum manner, so the batteries may have to be checked periodically to ensure they are not 'overcharged' or 'deep discharged', both of which can be damaging to the battery.

The next advance on this setup is to provide some form of regulation – this will ensure that the power supplied to the battery and the load are regulated. Above, we can see a regulator inserted between the solar array and the battery and load. This provides regulation of the power coming from the cell to improve the quality of power delivered and protect against abnormal conditions.

Now that DC appliances are hard to come by, some 12V appliances are being made for caravans and for use in cars. By using 12v DC direct from the solar array, you can avoid the losses of having to transform the power to high voltage AC using the inverter – which has inherent losses. Using 12v DC directly can be a more efficient use of electricity, where 12v equivalent appliances are available. However, most domestic users will want to use the same appliances as are commonly used in the home running from AC mains current. We saw in the preceding section, that inverters convert low-voltage DC to high-voltage AC, sacrificing current for voltage.

A stand-alone setup with a stand-alone inverter is much the same as the setup above, with a regulator controlling the supply from the solar array, a battery bank for storage, but with the addition of an inverter, and AC load. This setup gives you the flexibility still to use low voltage DC loads – remember, when you use the inverter to convert DC to AC, you lose power in the process as a result of the efficiency of the inverter. A block diagram of the layout is shown below.

One of the problems with this type of setup is that, although you have some battery storage, you are still beholden to the sun to provide your electricity – in the winter months you may find that you have insufficient power to see you through. You can complement your solar array with other renewables, such as micro-wind or micro-hydro, but one of the easiest ways of providing power-on-demand for an 'off-grid' system is to install a diesel generator. Even here, you still have many options as diesel engines, with minor conversion, run happily on either 'straight' vegetable oil or 'biodiesel', which can be made from vegetable oils. As these fuels are both derived from plant matter, unlike fossil-fuel diesel, they can be considered to be 'carbon neutral' and, as such, have a lower impact on the environment than burning fossil fuels. With a diesel generator, it makes more sense to power AC loads directly and charge the battery through a regulator, which converts the high-voltage DC to low-voltage AC, resulting in the setup illustrated on page 72. This system has a manual transfer switch in order to switch between AC loads powered by inverter or diesel generator.

Now we have covered off-grid systems, we will begin to look at grid-connected systems, which interact with the National Grid.

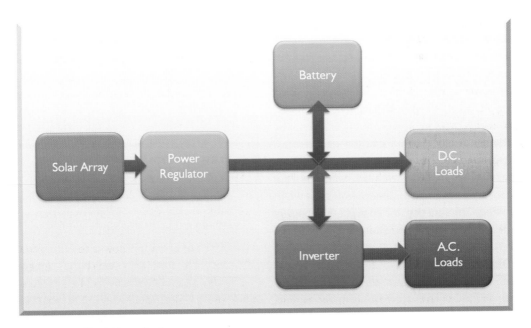

Basic AC/DC off-grid installation.

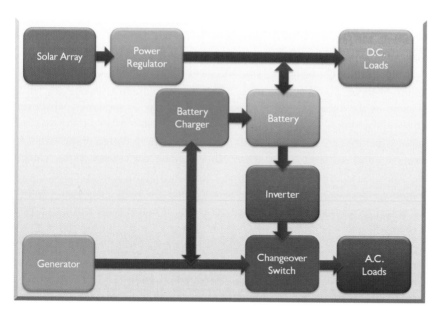

AC/DC setup augmented with generator.

Grid-Connected Systems

In the first system that we are going to look at, power comes 'from' the grid only – to augment the system when the solar panels cannot meet the on-site loads. This system retains the battery bank for storage capability – so, in the event of grid-failure, blackout or brownout, the site can run independently of the grid. The grid can provide power to the system in two ways: by powering AC loads directly, bypassing the inverter; and also by providing power to the DC circuit by a charger. A block diagram of the setup is illustrated opposite. This setup requires a multifunction inverter, in order to work with the grid when required and independently when required. The functions of charging the battery bank and regulation may be integrated into the inverter.

One of the drawbacks of this arrangement is that excess power cannot be sold back to the grid. There is a real satisfaction in receiving a cheque from your electricity supplier and knowing that you have helped to meet some of your neighbour's power demands with your clean, green power.

The next iteration of the grid-connected system does away with the battery and the 'DC infrastructure' and is the one that you are most likely to see in a UK home with a photovoltaic array connected to the grid. This system does away with 'independence' from the grid, as there is no on-site storage; however, the grid

IMPORT/EXPORT RENEWABLE ENERGY TARIFFS

It is highly probable that your household energy use will not be perfectly synchronized to the availability of solar energy, and so the excess energy produced in times of over-supply must either be stored or 'sold' back to the grid. There are different systems for doing this depending on the regulatory regime.

AC/DC setup with grid supplying supplementary power.

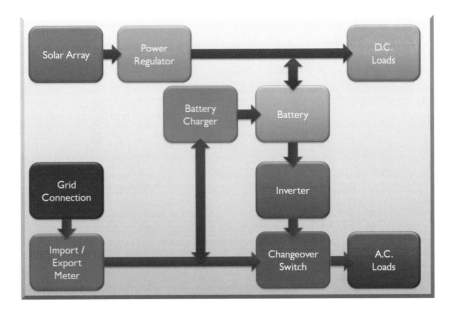

provides the role of the batteries in the system. When you produce power, but don't need it, the power is 'sold' back to the grid. When you need power but there is insufficient solar resource, you 'buy' power back from the grid. You may have a mechanical meter that runs 'backwards and forwards' or, if you have digital metering, you will have one meter for 'electricity in' and another for 'electricity out' or even a sophisticated digital meter that integrates the functions of 'net metering' into one meter.

NET METERING

Net metering is the process by which an electricity supplier takes the amount of energy you have produced by your solar array and subtracts it from the amount of energy used from the grid. This works in favour of the consumer because you get paid the same price for electricity you generate, as the electricity you use. The alternative is net purchase and sale, where you, as the consumer, end up being paid the wholesale 'buying price' of electricity, which is much lower than the price you would normally pay to buy a unit – a raw deal. Check with your supplier to see what arrangement they are prepared to offer you.

DUAL METERING/AVOIDED COST

Under this regime, the worst for producers of clean renewable energy, the supplier will only pay you for the cost they have 'avoided' in not having to produce the energy you have supplied to the grid. The value that is attributed to the energy varies widely depending on the formula used by the utility company to calculate the avoided cost.

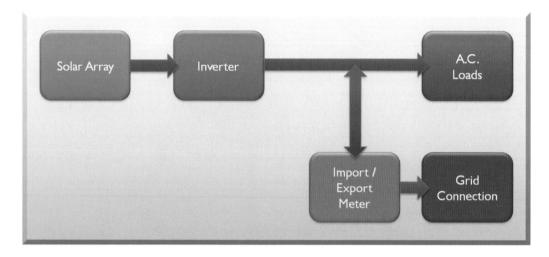

AC/DC grid synchronous setup.

One of the advantages of an AC/DC grid synchronous setup is that you do not need a solar array from the outset that will meet your full needs – as you can choose for solar power to provide only a 'proportion' of your electricity needs. This is good news for people with small roofs who do not have big gardens or other areas to accommodate their solar arrays!

For this type of installation, you will need a 'synchronous' inverter, which synchronizes the waveform of the sine wave the inverter produces, with that coming from the grid. This is very important, as two sine waves 'in phase' will reinforce each other; however, if the waves were 'out of phase' one would be working against the other, and the sine waves would be trying to cancel each other out.

CHAPTER 5

SOLAR WATER HEATING

Stages in Designing a Solar Hot-Water System

The process of choosing a suitable solar hot-water system is a simple one, if the steps are followed in the right order. By the end of this chapter, you should have a firm grasp of the technologies that can be employed successfully in the UK, to heat your water using the sun. However, it is also important to have an understanding of the correct process to follow, in order to select appropriate components for your system. It is beyond the scope of this book to provide detailed instructions on installation (there are some good books in Further Reading that will help you expand your horizons in this respect); neither is it possible within these pages to provide details of all the calculations required for every variant of system. However, it is hoped that by the end of reading this chapter, you will be an informed consumer, able to make well-reasoned choices about the appropriate selection of equipment for your application.

Step 1: Sizing the Storage-Cylinder Capacity

If you read the section on conducting a home water-audit, you should have a firm grasp of your household's domestic hot-water requirements and usage patterns.

Step 2: Selecting the 'Type' of Storage Cylinder

I include a thorough review of domestic hot-water storage-cylinder technologies, and by the end of reading this section, you should have a firm grasp of what assets you already have within your home, their potential for use and other technologies that you should consider for upgrading your hot-water storage capacity.

Step 3: Selecting the Collector Type

There is a review of the different collector technologies that are suitable for use in the UK. Please be aware when reading this section and comparing to other sources of information (the internet and international publications) that there are other collector types available, but that they may not necessarily be suited for UK use. There is a review of the different collector technologies that are suitable for use in the UK, and other locations with similar weather conditions; note that there are other collector types available, but they may not neccessarily be suited for use in countries like the UK, with its relatively cold climate.

Step 4: Selecting the Collector Position

The earlier sections of this book should provide

an invaluable guide to making optimal use of the solar resource, and some of the notes on collector positioning for photovoltaic devices are equally applicable to solar thermal installations. Be aware that you may be limited in your positioning options by planning permission, if you opt for a configuration other than having your collector parallel to your roof surface, or if you live in a conservation area or area of outstanding natural beauty. For a fixed-system configuration in the UK, it is ideal if it is due south-facing and inclined at 35 degrees to the horizontal for all-year-round performance. However, a steeper inclination will optimize the system for autumn and spring performance at the expense of spare capacity in the summer. Bear in mind that orientations between south-east and south-west will also perform without a significant degradation in performance, and that inclinations of between 15 and 50 degrees will yield favourable results.

Step 5: Selecting the System Configuration

The later sections of this chapter examine how these components connect together and the different solar thermal configurations that are available. If you have existing components of a central heating system, your priority may be how to integrate solar with your existing heating system effectively, with the minimum of new components – or your priority might be the most efficient system, designed from the ground up. Either way, the configurations in this chapter cover most setups suitable for UK solar installations.

Step 6: Select Pipe Sizing and Insulation

It is impossible to make broad recommendations without consulting the specific nature of a solar installation. Furthermore, this is the sort of judgement that requires the knowledge of a qualified heating engineer, and is beyond the scope of this book for homeowners and builders. However, there are a number of factors to consider.

Larger diameter bore pipework will allow water to flow freely but there is a trade-off in that the increased diameter of the pipe allows a greater amount of heat to be lost. Smaller diameter pipes will have a smaller overall surface area, but because it's harder to force water through a narrower pipe, it will introduce a parasitic load into the system, which will require additional pumping in order to move the water around the system. It goes without saying that whether you select larger or smaller pipework, ensure that it is well insulated to prevent valuable heat that you have generated from being lost.

Manufacturers and resellers of solar domestic hot-water heating components should be able to make recommendations on what plumbing fittings and sizes are suitable for use with their products.

Step 7: Select Pump Size

The pump circulates the working fluid around the solar hot-water heating system, moving the fluid (and the heat with it) from collector to storage tank and back again. It is important to ensure that the pump is correctly sized to ensure adequate performance of the system. However, an over-specified pump will consume excess electricity and introduce an efficiency penalty in the energy-balance of the system. Again, this is the sort of decision to be taken by a professional and beyond the scope of this book.

Step 8: Select Solar Controller

The solar controller will monitor the temperature

at a number of points in the system and select when it is appropriate for the working fluid to be circulated. Circulating the fluid unnecessarily will waste electrical power to drive the pump and, furthermore, may be detrimental to performance, as driving hot water through a cold collector will heat the roof and not a lot else.

Some systems, such as the Solartwin system, are effectively self-governing. The irradiance on the solar photovoltaic panel, which drives the pump for the system, governs the speed at which the pump is driven – and because there is no import/export of electricity to the main system (the wiring being self-contained to the system), there is no risk of wasting energy.

Hot Water Total Daily Consumption			
Type of Use	Daily Average Use	Litres	Daily Average Use × Litres =
Washing (Personal)			
Frugal Bath		30	....
Indulgent Bath		90	....
Shower (Mains Pressure)		15	....
Shower (Power)		50	....
Washing (Items)			
Washing-Up By Hand		5	....
Dishwasher		25	....
Washing Machine		80	....
Miscellaneous			
....		....	....
....		....	....
....		....	....
Total Daily Consumption			(I)
Utilization Factors (II)			
Hot Water Used When Available Throughout Day	0.8		
Low Hot-Water Use Spread Through The Day	0.9		
Average Hot-Water Use Spread Through The Day	1.0		
Washing-Machine Used During The Afternoon	1.3		
High Morning Usage – Low Usage During Daytime	1.5		
Minimum Hot-Water Cylinder Storage Size (III)			
Minimum Cylinder Size (III) = Total Daily Consumption (I) × Utilization Factor (II)			
....................(III) = ...(I) × ....(II)			

Conduct a Domestic Hot-Water Audit

Before beginning to look at installing solar water heating, you should get to know the demand in your household for hot water. You can do this by conducting a home audit – get your family to keep a diary of how much water they use over the period of a week. Stick post-it notes near the bath, shower, sink, dishwasher, washing-machine and any other items that use hot water, and get your family (and even your guests) to keep a tally of how they consume hot water. Draw a line under each day and calculate the average usage every day, then compute the data into the table on page 77.

How Can I Change My Hot-Water Usage Behaviour?

We all love a deep bath now and again; however, by adjusting our behaviour, we can be more frugal with our energy needs, which means we use less energy to supplement our solar system. Think of your lifestyle and how you could attempt to change it to use less hot water. Also, are you the sort of person who gets really frustrated if there isn't hot water 'on demand' or are you the sort of person who can adjust your lifestyle and routines to when freely heated solar hot water is available? The bottom line is both economic and environmental.

Calculating Your Daily Requirement for Hot Water

First transfer your tally to the Daily Average Use column, then for every line, multiply the number of times you use the item by the amount of water it consumes each time you use it over the course of a day to give your daily use line totals. These line totals are useful in helping you to diagnose where most of your household hot water is being used. In the table, there are some suggested figures; however, if you have data that your appliance consumes a certain amount of water, each use, do not hesitate to use that instead.

Now, different people use their domestic hot water in different ways, and there are different patterns of usage. We need to compensate for these different behaviour patterns by using a Utilization Factor. This factor allows us to adjust for the different ways people use their hot-water supplies. Of course, the higher the utilization factor, the greater the size of cylinder you will need for storage – so it is cheaper to adjust your behaviour than to fit a larger cylinder!

How Do I Compare to the Average?

[The average person uses] 31 litres of hot water per week for washing machines, 8 litres for dishwashers, 85 litres for showers and 158 litres for baths. If we include an allowance for washing up at the sink by hand this rises by a further 38 litres per person per week. Finally we can allow an additional 35 litres of hot water per person per week for hand and face washing. This gives a total average use of around 355 litres of hot water per person per week.

J. Hulme (2005) 'Domestic Hot Water Use In England', Building Research Establishment

Now you've got some idea of the size of cylinder you require, let's evaluate the different kinds of thermal store cylinder available for your installation.

Construction of a copper hot-water cylinder.

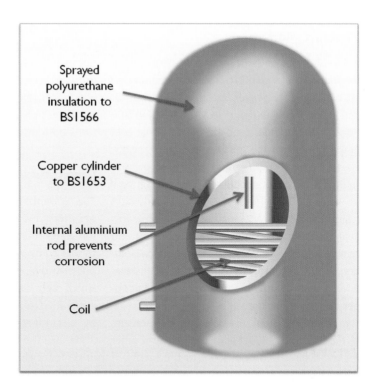

Sprayed polyurethane insulation to BS1566

Copper cylinder to BS1653

Internal aluminium rod prevents corrosion

Coil

Solar Thermal Cylinders

The solar collector takes cold water and heats it using the Sun's energy in order to produce hot water. However, the collector is not the end of the story. For a direct system, a hot-water store or cylinder is required in order to provide some buffering between supply and demand; whilst for an indirect system, the cylinder serves the dual purposes of storing the heated water and acting as containment for a coil, which acts as the heat-exchanger between the collector panel sealed-circuit and the household DHW system.

Copper cylinders are a proven solution, which have been used to store hot water in UK homes for around a century. However, in some hard-water areas, limescale build-up can produce a problem, and so stainless steel cylinders are also available. The construction of a copper hot-water cylinder is shown above. A copper cylinder consists of a copper sheet manufactured to BS EN 1653 specification, which is either brazed or butt welded to form the containment for the hot water. The thickness of the sheet is determined by the pressure that the cylinder must operate at. The cylinder is tested at 1.5 times the working pressure that it needs to operate at. An piece of aluminium is fitted to the inside of the cylinder – copper does not corrode particularly easily, however, in the event that any corrosion does take place, the aluminium will corrode in place of the copper, sacrificing itself rather than the cylinder deteriorating.

Cylinders are then coated with an externally applied insulation – usually a polyurethane foam. BS1566: Part 1 specifies that the heat loss from a hot-water cylinder must be no more than 1W for every litre of cylinder capacity.

Vented Cylinders

The vented hot-water cylinder is probably one of the most popular solutions you will encounter in UK plumbing – however, it is peculiar to the UK and Ireland and is much less popular internationally. The cylinder requires an additional cold tank (the tank in the loft) located some height above the highest tap outlets in order to provide sufficient head to produce a reasonable flow of water at the highest outlet. The cold-water header tank can produce a number of problems (leaking, freezing, contamination with dust and *Legionella*), but because it is an economical solution, it has become very popular in UK installations. Commonly, cylinders are available in copper in a variety of capacities.

In a vented system, the flow rate and pressure of the hot water available at the taps and other outlets is directly proportional to the height of the cold-water header tank, and hence the amount of head it can produce. One of the caveats of a vented cylinder is that, if multiple hot-water outlets are used at the same time, a drop in pressure and performance can result, which can be irritating for some users, for example, poor flow of hot water when trying to run a bath or whilst the washing machine is on.

Unvented Cylinders

The unvented cylinder is a solution that is becoming increasingly popular in UK domestic hot-water installations. An unvented cylinder should come with safety devices and controls to prevent over-temperature and over-pressure, which can cause a dangerous situation. It is imperative that an unvented cylinder is installed by a competent person who is qualified to undertake the task, as incorrect installation of an unvented cylinder could result in injury and or death.

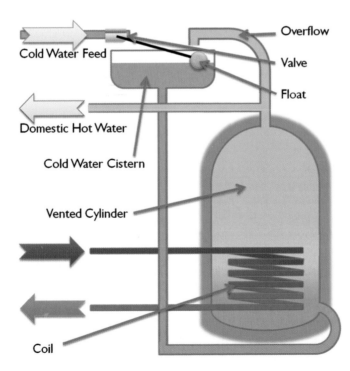

Cold Water Feed

Domestic Hot Water

Cold Water Cistern

Vented Cylinder

Coil

Overflow

Valve

Float

Vented hot-water cylinder.

Unvented hot-water cylinder.

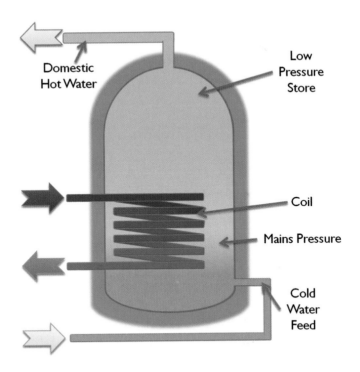

Because of the increased pressures, unvented cylinders must be made from a heavier gauge of metal and also to higher tolerances – meaning that they are more expensive than vented cylinders. There is also less of a selection of vented cylinders available than unvented cylinders.

Unvented cylinders have grown in popularity in the UK, with homeowners' desire to remove header tanks from the loft, creating a clear space that can be converted for domestic use.

Unvented cylinders can operate at pressures of up to 3bar, providing flow rates in excess of 25ltr/min. This means that water can be drawn off from one outlet at the BS7000 specification of 18ltr/min, whilst still having additional capacity to operate other outlets. This ability to supply more than one outlet at one time can be attractive to some consumers, especially those whose system will have a high rate of usage and where a drop in pressure when a number of outlets are used simultaneously would be undesirable.

Cylinder with Integral Coil

It is possible to combine the safety benefits of a cylinder operating at a low pressure with the advantage of having hot water available at a high pressure. A cylinder with integral coil stores the thermal energy of the solar system in the mass of the water contained in the cylinder. However, within this cylinder is a coil containing water at high (main's) pressure, which is heated by the surrounding hot water in the cylinder. This arrangement has the advantage that no large 'tanks' are required, but as a result of the elimination of tanks, there is no backup if the main's water supply fails.

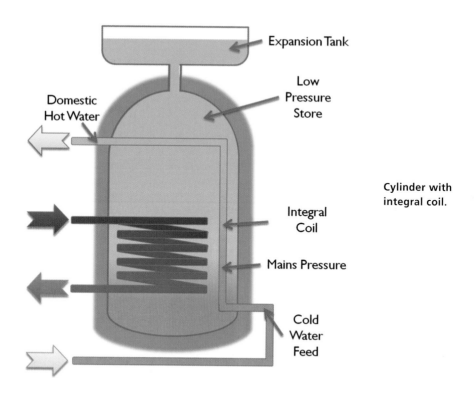

Expansion Tank

Low Pressure Store

Domestic Hot Water

Integral Coil

Mains Pressure

Cold Water Feed

Cylinder with integral coil.

Cylinder with external heat exchanger.

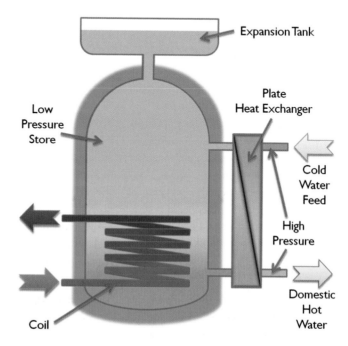

Expansion Tank

Low Pressure Store

Plate Heat Exchanger

Cold Water Feed

High Pressure

Domestic Hot Water

Coil

Cylinder with External Heat Exchanger

Another alternative is to use an external heat exchanger with the cylinder, if the cold water feed is of a sufficient pressure. The water on the cylinder side of the heat exchanger is all at low pressure, whilst the water on the domestic, hot-water side of the heat exchanger is at mains pressure. For times of large hot-water consumption, such as running a bath, there will need to be high temperatures inside the cylinder to ensure an appropriate quality of service. In this system, hot water is not being 'stored' – the cylinder is acting as a store for heat not water. The heat exchanger effectively 'produces' hot water on demand and the temperature will largely depend on the amount of available heat within the thermal store of the cylinder.

Because the heat exchanger is an external component, it can be removed for periodic cleaning and maintenance, allowing the build up of limescale to be kept under control in hard water areas. Again, there are no tanks with this particular solution, so there is no reserve supply of water in the event of failure of the main's supply.

The IDM Hygenik Tank is an example of a thermal store with an external heat exchanger. Another purported benefit of this design is that, rather than drawing water from a tank, where sludge, limescale and debris can accumulate in the bottom over time, water is only passing through the heat exchanger, which has a smaller area and can also be removed for periodic maintenance and cleaning.

IDM Hygenik Tank. (Invisible Heating)

Retrofit Heat Exchanger

It is possible to retrofit a heat exchanger to your existing hot-water cylinder in place of the immersion heater. The retrofit heat exchanger is a coil or straight section of copper tube, with a mounting and flanges the same dimension as a standard immersion heater. A retrofit solution will work best if your immersion heater is located in the bottom of your cylinder. This is unfortunate, as many cylinders have a top-mounting immersion heater, which makes for poor efficiency with a solar water-heating system. Whilst this is a vastly inferior option, it is a cheap solution, which allows you to stagger the cost of fitting solar collectors and allows you to pay for an upgraded hot-water cylinder or thermal store at a later date.

Twin-Coil Solar Thermal Store

In the best installations, a twin-coil cylinder will be employed. With a twin-coil cylinder, there is the advantage that for a given quantity of water, the surface area subject to heat loss will be smaller with a twin-coil cylinder than with a twin-cylinder system of equivalent volume.

The solar collector connects to the bottom coil, and the supplementary boiler connects to the upper coil. As the solar collector heats the warm water, it will rise within the cylinder. A sensor within the cylinder will detect whether supplementary heat is required and augment the

heat from the solar collector, if necessary. This design results in a tall tank, which is good from the point of view of thermal stratification. Furthermore, because the supplementary heat is integrated into the same tank as the solar collector, no supplementary heat source is required for sterilization of the tank. Hot water is drawn from the top of the tank, with cold water being introduced at the bottom to replace the drawn-off hot water. Because the solar coil is at the bottom of the cylinder, it acts on fresh, cold water directly, maintaining a differential between the temperature of the cylinder water at the bottom of the tank and the temperature of the solar coil. This is the key to efficient heat transfer.

There are relatively few disadvantages to this particular arrangement – there may be some additional expense incurred for the specialist cylinder, and the cylinders are quite large, which may preclude installation in certain circumstances; however, the performance of this solution is very good.

Twin Cylinder Arrangement With Pre-Heat Cylinder

Two separate cylinders can be used where the first cylinder (heated by the solar collector) is a pre-heat cylinder, which provides warm water to the second cylinder, where it is brought up to temperature, if necessary, by supplementary heating. As hot water is consumed from the main cylinder, water flows from the pre-heat cylinder, where extra heat can be added, if necessary, by a supplementary boiler. There is a caveat with this system, that if hot water is not being drawn off, solar heat can become 'caught' in the pre-heat cylinder – resulting in the main cylinder being heated by conventional means anyway.

This method permits easy conversion of existing installations and also allows for storing large

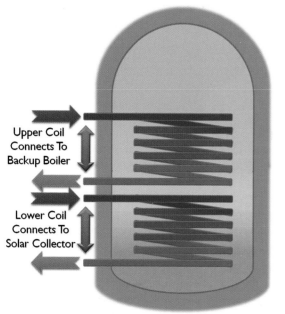

Upper Coil
Connects To
Backup Boiler

Lower Coil
Connects To
Solar Collector

Twin-coil solar thermal store.

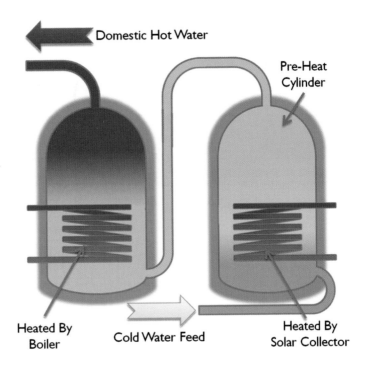

Twin cylinder arrangement with pre-heat cylinder.

Domestic Hot Water

Pre-Heat Cylinder

Heated By Boiler

Cold Water Feed

Heated By Solar Collector

quantities of water. For safety, both tanks must be fitted with a pressure relief header tank allowing for expansion. Furthermore, as the pre-heat cylinder may not always reach a sufficient temperature to enable sterilization, an additional heat source (e.g. immersion heater) should be fitted to ensure that the tank can periodically be brought up to temperature and sterilized.

Cylinder Storage Ratio

The ratio of hot water delivered over the size of the cylinder capacity is an essential concept to grasp to ensure system efficiency. In a well-designed efficient solar system, there will be a dedicated capacity for 'pre-heating' by the solar system. This could be a separate pre-heat cylinder, which feeds into the main system, or it could be the proportion of a thermally stratified tank dedicated to pre-heat capacity.

In a single tank, an invisible dividing line can be drawn between the top section, which is used for storage of 'hot' water ready for draw-off, and the bottom half, which contains colder water for pre-heating to ensure effective heat-transfer from the solar collector.

Ensuring Thermal Stratification

The performance of a solar installation will be improved if 'stratification' can be created within the solar cylinder. Stratification is the process whereby less dense hot water rises to the top of a cylinder, and cold water sinks to the bottom. A tall cylinder will aid stratification more than a short, squat cylinder.

By creating stratification within a solar cylinder, it is possible to extract hot water at the top, whilst replacing cold water at the bottom of the tank. We want to ensure that the 'hot' water is as

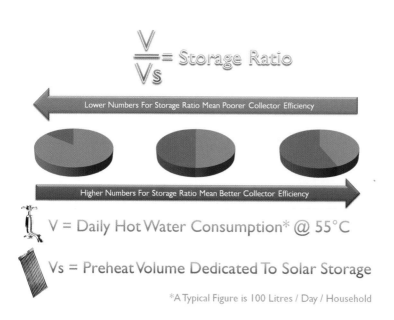

$$\frac{V}{V_s} = \text{Storage Ratio}$$

← Lower Numbers For Storage Ratio Mean Poorer Collector Efficiency

Higher Numbers For Storage Ratio Mean Better Collector Efficiency →

V = Daily Hot Water Consumption* @ 55°C

V_s = Preheat Volume Dedicated To Solar Storage

*A Typical Figure is 100 Litres / Day / Household

LEFT Storage ratio is crucial to performance.

BELOW Graph showing solar storage efficiency depends on preheat volume ratio (data sourced from BS5918).

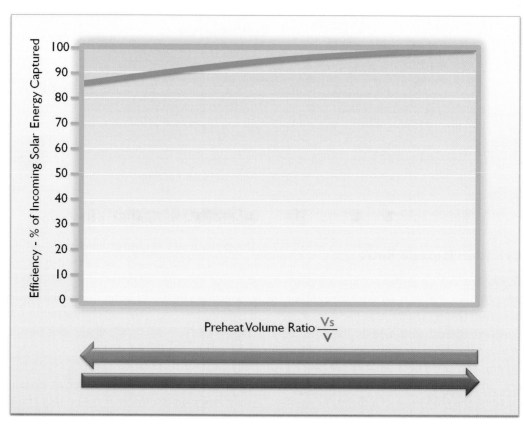

Efficiency - % of Incoming Solar Energy Captured

100 90 80 70 60 50 40 30 20 10 0

Preheat Volume Ratio $\frac{V_s}{V}$

RIGHT **The aspect ratio of the tank is crucial to achieving good thermal stratification.**

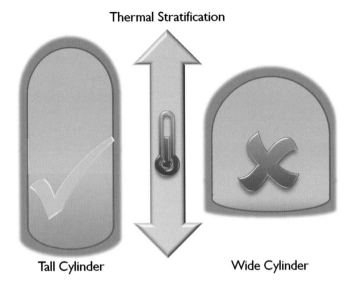

Thermal Stratification

Tall Cylinder   Wide Cylinder

BELOW **Proclean stratified storage tank. (Invisible Heating)**

hot as possible, for quality of service; however, for efficient heat transfer, we want to ensure that the water at the bottom of the cylinder is as cold as possible. Stratification allows us to have both hot water and cold water within the same tank, albeit separated into layers with a thermal gradient.

Stratification can also be aided by internal baffles, ensuring that the force of the cold water entering the cylinder does not induce swirl, mixing the hot and cold water within the tank.

Solar Thermal Collectors

Solar Thermal Collectors – Then and Now...

With the increasing focus on renewable energy technologies, investment into research and development, and increasing consumer interest, manufacturers of solar thermal collectors are beginning to realize the economies of scale that come with manufacturing solar thermal collectors in bulk. In the past decade, we have seen solar collectors move from being an esoteric, exotic item, to something that is firmly rooted in the nation's psyche, and increasingly being sold through 'traditional' building material outlets and plumber's merchants, rather than just through specialist suppliers. A couple of decades ago these economies of scale were not there, solar collectors were expensive and so resourceful experimenters turned to building their own. A relatively simple solar collector can be made from a radiator, enclosed in an insulated, glazed housing, painted black. However, the performance and response of such a collector is inferior to commercially produced units, which are made to high specifications.

In this chapter, we will be focusing on high-quality, commercially produced solar collectors. For your installation to be grant-funded, you will need to obtain your materials from a supplier that meets the bench-marks that are in force at the time of installation, and your installation will need to be installed by a certified installer. In this respect, a DIY panel and installation would not be suitable.

However, DIY construction and experimentation can still provide fruitful outcomes and should not be neglected as an appropriate technology is some circumstances. It is beyond the remit of this book to provide a thorough discussion of the subject, so please see Further Reading.

How Much Energy Do Solar Thermal Collectors Provide?

The efficiency of a solar thermal collector, which we will now call the 'collector efficiency', is expressed as the amount of solar energy that 'hits' the solar collector and that can be converted into useful, usable energy.

For domestic hot-water applications in the UK, the only collectors worth considering are some sort of glazed or insulated solar collector – whether that be flat-plate or evacuated tube. An unglazed collector will provide low-temperature hot-water in the summer, for applications like swimming pools; however, because of the higher target temperatures, they are unsuitable for domestic hot-water solar collectors.

We have thus ascertained that, before any 'collecting' is done, our solar energy must first pass through a layer of glass – whether this be the flat glass of a flat-plate collector, or the cylindrical glass of an evacuated tube. When solar radiation hits glass, a proportion of it is reflected away and a proportion passes through the glass to our collector. A small component of the solar radiation is also absorbed by the glass and some of this is radiated to the atmosphere and some to the collector – this is known as secondary heat

**Optical and
thermal losses
and usable heat.**

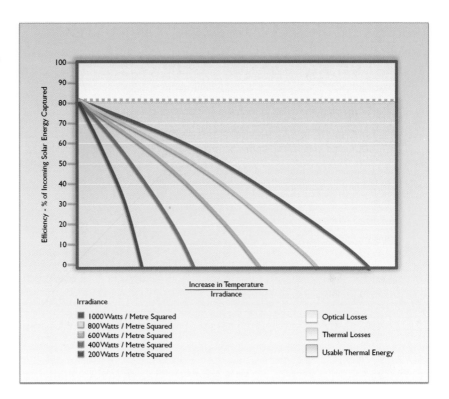

gain. So, there is some inherent 'optical ineffi-ciency' in a solar collector, as a result of the need to pass through a layer of glass. There is also the ability of a solar collector to 'absorb' the solar energy, which will to a large degree depend on the optical coating of the solar collector.

Now, to thermal efficiency. A solar collector is an insulated box – and whilst there will be some heat produced by the absorbed solar energy, there will also be some heat loss, as a result of this heat being transferred into the cold environ-ment (your roof) in which the solar collector operates. This problem is exacerbated by the temperature difference between the collector surface and the environment in which it oper-ates. Simply put, the colder it is outside for any given temperature, the less efficient your collec-tor will be at any given insolation (exposure to

the Sun's rays). This thermal loss is also known as the 'k-value'. You want to look for a solar collec-tor with the highest possible conversion factor and the lowest possible k-value. The efficiency of a solar collector can thus be expressed by a graph as shown above.

The 'optical losses' of the collector are fixed, whereas the 'thermal losses' depend on insola-tion and the temperature differential with the usable, harnessed energy highlighted red, shown beneath the lines. The 'optical' loss defines where the line intercepts the y-axis of the graph, whilst the thermal losses affect the slope of the line. We can use graphs like this to compare solar collectors for different applications.

The graphs on the next page show the regimes under which collectors operate in the

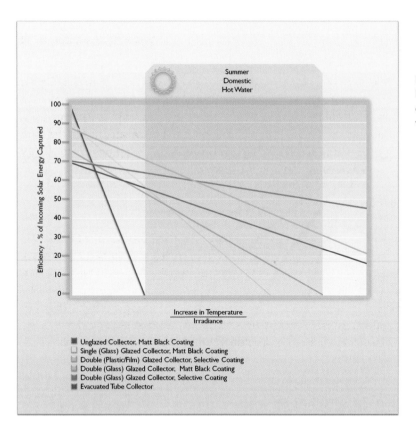

Performance band for summer domestic hot water.

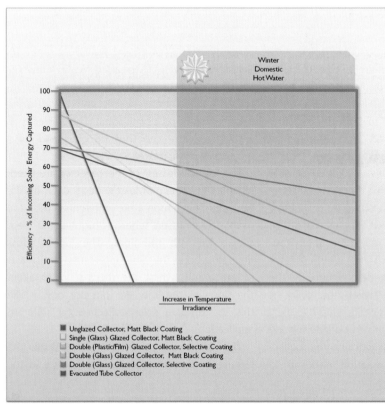

Performance band for winter domestic hot water.

summer and the winter. We can use these graphs to examine the performance of different solar collectors under a range of conditions. Whilst unglazed collectors appear to offer a good performance to the left-hand side of the graph, they are only really suitable where the ratio between the increase in temperature required over the irradiance is small.

Absorber Surfaces for Solar Thermal Energy

The simplest surface we can use to absorb solar energy is a flat, black-painted surface. Black absorbs all frequencies and reflects none, so the incoming solar radiation will be absorbed by the surface and cause it to get hot. However, there is one small caveat here – once the temperature of the absorber rises above the ambient temperature, the surface will re-radiate heat in the form of long-wave radiation, losing some of its energy in the process.

The Flat-Plate Collector

The flat-plate collector consists of a flat metallic plate, which is coated with either a matt-black or a selective coating. A matt-black coating absorbs a high proportion of the incoming solar radiation and heats up as a result. The only caveat is that, as the plate heats up, it re-radiates a proportion of the solar energy absorbed.

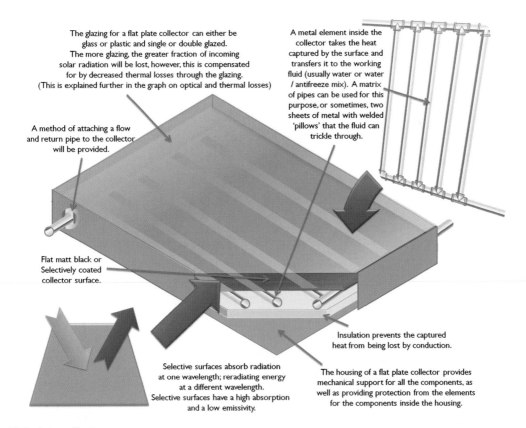

The glazing for a flat plate collector can either be glass or plastic and single or double glazed. The more glazing, the greater fraction of incoming solar radiation will be lost, however, this is compensated for by decreased thermal losses through the glazing. (This is explained further in the graph on optical and thermal losses)

A metal element inside the collector takes the heat captured by the surface and transfers it to the working fluid (usually water or water / antifreeze mix). A matrix of pipes can be used for this purpose, or sometimes, two sheets of metal with welded 'pillows' that the fluid can trickle through.

A method of attaching a flow and return pipe to the collector will be provided.

Flat matt black or Selectively coated collector surface.

Insulation prevents the captured heat from being lost by conduction.

Selective surfaces absorb radiation at one wavelength; reradiating energy at a different wavelength. Selective surfaces have a high absorption and a low emissivity.

The housing of a flat plate collector provides mechanical support for all the components, as well as providing protection from the elements for the components inside the housing.

Flat-plate collector.

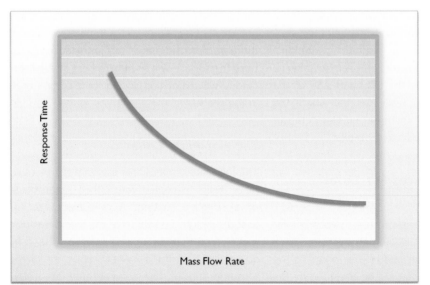

Response Time

Mass Flow Rate

LEFT **Response characteristic of flat-plate collectors in changing irradiance (data from BS5918).**

BELOW **Flat-plate collector on the roof of property in Freibad, Germany. (Maria Hawton-Mead)**

Selective surfaces overcome this problem; they are engineered so that the characteristics for absorbing radiation and re-radiating radiation are different.

The plate can be coupled to a matrix of tubes that allow a transfer fluid to flow underneath the flat plate, taking the heat from the plate away to the heating system. The alternative is for two sheets of metal to be brought together in close proximity, with a number of small welds forming a 'pillow' structure, which allows the transfer fluid to trickle through. The flat plate and matrix are contained within a housing, which provides mechanical support for all components, and will have fixings to allow mounting to the roof. The housing will be well insulated on the surfaces under the flat plate and matrix, and also on the sides of the box.

Unglazed collectors, which are simpler in construction, are only used for swimming-pool applications. For domestic solar water heating, some glazing is often employed to prevent conductive heat-loss from the flat plate back to the atmosphere. The glazing can either be single- or double-glazed, and made of either glass or plastic. The use of glazing does result in a small 'optical loss' – some of the solar energy will be lost as a result of being absorbed and/or reflected; however, this is more than compensated for by the improvement in insulation that the glazing affords.

A DIY approach often used in the past, was to use black-painted recycled radiators in a housing as a simple solar collector. Whilst this design can provide a cheap collector, the high volume of water within the collector means that it is not

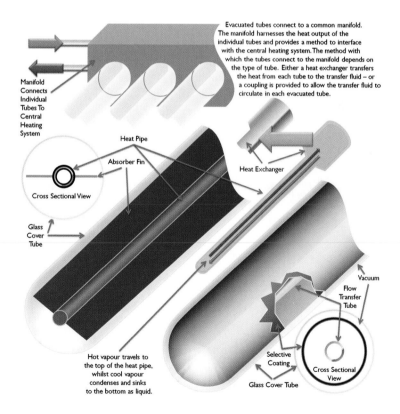

Evacuated tubes connect to a common manifold. The manifold harnesses the heat output of the individual tubes and provides a method to interface with the central heating system. The method with which the tubes connect to the manifold depends on the type of tube. Either a heat exchanger transfers the heat from each tube to the transfer fluid – or a coupling is provided to allow the transfer fluid to circulate in each evacuated tube.

Manifold Connects Individual Tubes To Central Heating System

Heat Pipe

Absorber Fin

Cross Sectional View

Heat Exchanger

Glass Cover Tube

Vacuum

Flow Transfer Tube

Hot vapour travels to the top of the heat pipe, whilst cool vapour condenses and sinks to the bottom as liquid.

Selective Coating

Glass Cover Tube

Cross Sectional View

Cutaway evacuated tubes.

particularly responsive to changing conditions. We can see in the graph on page 92 how the response of a solar collector is related to the mass flow rate of the collector – the less water the collector 'stores', the more responsive it is.

The Evacuated Tube Collector

Evacuated tubes have a more complicated construction than flat panels – as a result of this, they are commensurately more expensive. Evacuated tubes are used in groups, which are connected at one end via a common manifold. Inside the glass tube, a vacuum is created to minimize heat losses – it does this by virtually eliminating losses as a result of conduction or convection.

There is a small point of difference here, in how the heat is conducted to the manifold – this gives rise to two distinctly different kinds of evacuated tube collector, which we will now discuss: the direct flow tube collector and the evacuated heat pipe collector, which are contrasted in the image on page 93.

Direct Flow Evacuated Tube Collectors

In a direct flow evacuated tube collector, the working fluid (the water in your heating system) actually flows through the pipe in the evacuated tube collector, heating the water directly and returning it to the manifold, where it is connected to your hot-water system.

The evacuated tube collectors of this type comprise a fin housed within an evacuated glass tube or a circular collector, with concentric tubes

LEFT **Cutaway evacuated tube – direct flow.**

RIGHT **End of direct flow evacuated tube – detail.**

Direct flow evacuated tube – manifold detail. (Schott UK)

providing flow and return. The fin will have a flow and return pipe integrated within its construction. At the top of the tube, an interface is provided to allow the tube to mate with the manifold, which couples the array of tubes together, and takes the flow and return from the heating system and couples it to the individual flow and return connections on the tubes.

Evacuated Heat Pipe Collector
The evacuated heat pipe collector is subtly different, in that the working fluid itself does not pass from the manifold into the evacuated tube collector. Instead, a sealed copper pipe inside the evacuated tube, containing another transfer fluid (such as alcohol), heats up and, as it does

so, produces a hot vapour, which rises through the tube to the manifold. As the manifold is cooled by the water that is being heated, the alcohol loses its heat energy to the water and condenses, returning to the bottom of the evacuated tube, where it is heated again.

A Word on the Transportation of Evacuated Tubes
Evacuated tubes are fragile creatures, being largely made of glass, with an internal vacuum; the tubes are very susceptible to damage. When transporting evacuated tubes, ensure that you take the utmost care, use correct packaging and don't drop them. Ensure that any couriers or transport services that you might use to move the tubes are explicitly aware of their fragile nature – if not, damage can easily result.

Installation of Solar Collectors

Solar collectors are most often roof-mounted, requiring working at height. If you are going to attempt your own solar installation, do take appropriate precautions – ensure the roof surface on which you are working can support your weight, or use suitable measures to spread the load. Also, if you are working at height, ensure that you have someone on the ground to help watch and ensure that, if you have an accident, there is help on hand. The person on the ground can also be helpful for lifting items in a safe manner. When you penetrate the tiles or roof surface, make sure that the penetration is properly insulated and flashed to prevent ingress of water and to ensure that no heat is needlessly lost from the pipes in this area.

When siting your solar collector, you might also want to consider access. If your roof forms part of a room or a loft conversion, you could site your solar collector in such a manner that access is easy from adjacent windows or access

LEFT Take care in the transportation of evacuated tubes. (Solar UK)

BELOW Installing evacuated tubes – take care when working at height. (Schott UK)

Evacuated tube collectors on a family home. (Schott Ltd)

hatches. In the illustration above, the solar collector has been installed below a large window – this has the dual benefits of affording easy access for installation and maintenance, whilst also giving a clear view of the collector from the inside of the house; ideal when you want to show people your installation.

Mounting Options

In-Roof

It is possible to achieve a flush installation with some types of solar thermal collector. Such panels will be supplied with a 'flashing kit', which allows the panel to be mounted and then a water-tight seal to be achieved between the panel and surrounding tiles. Additionally, some manufacturers are beginning to produce roof-integrated solar thermal systems, which will integrate with the tiling systems of some manufacturers. An example would be the C21 solar thermal slate from Solar Century, where the tiles will integrate with roof tiles, and even solar photovoltaic tiles.

On-Roof

The most common installation method with solar thermal installations is to have an 'on-roof' installation, where the collector sits above the roof line with an air-gap between the collector and tiles. This is a slightly more visually intrusive method of mounting, as the pipework, to and from the solar collector, is visible, as is the bracketing and other mountings. However, this is a

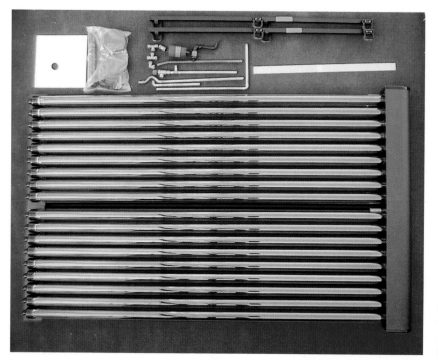

simple form of installation to retrofit to an existing roof. Your solar collector should come with a range of mounting hardware to permit on-roof installation.

Tracker

Another more sophisticated option, which may be suitable for some installations, is the construction of a solar tracker. Solar trackers follow the path of the sun, by changing the orientation of the solar collector to ensure that the collector is facing the sun at all times. The benefits from improved efficiency must be offset against the amount of energy needed to continually re-orient the solar panel. An array of trackers can be seen in this SolarUK installation, on the flat roof of these flats, where they are being used to supply domestic hot water for the occupants.

Evacuated tube trackers on roof of flats – detail. (Solar UK)

Solar Water-Heating Systems Terminology

Direct and Indirect Systems

There is a distinction to be made between direct and indirect solar hot-water heating systems. In the UK, indirect systems are by far much more common than direct systems.

Direct System

In a direct system, the fluid that flows through the solar thermal collector is the actual hot water, which will eventually make its way to your tank and, ultimately, to your tap.

Indirect System

In an indirect system, the circuit through which the water flows through your collector is physically separated from your hot-water cylinder. This can allow different working fluids to be used in the system; for example, a mixture of water and anti-freeze.

Freeze Protection for Your System

Water expands when it freezes, which isn't good news for solar collectors. A solar collector full of water in a cold environment can be a recipe for expensive repairs or even replacement. There are a number of strategies that we can use to help prevent damage to our solar collectors in this situation, and we will discuss these here.

Drainback and Automatic Drain Down

Drainback is used to prevent water from being left in the solar collector during cold periods, where the water could freeze, expand and damage the solar collector. A temperature sensor is coupled to the pump. A special drainback tank is used, which allows the fluid from within the solar collector to empty out, under the influence of gravity, and fill the drainback tank. When operation is resumed, the water from the tank is pumped back round the solar collector. Careful design of the drop is necessary to ensure that water drains out of the system correctly, and that no water is left. A variation on this is automatic drain down, which involves fluid being 'dumped' out of the system rather than flowing into a vessel when a freezing condition is near to being reached. This requires a manual refill of the system and air bleed – as such, it is not suitable for environs where a freezing condition is a regular occurrence, being better suited for warmer climes where cold weather is rare.

Freeze Tolerance

By using materials for pipework with a little bit of extra 'give', it is possible to make systems that are tolerant to freezing. This is a system sometimes employed in uninsulated, unglazed swimming-pool collectors, which are outside all day. As they are made out of flexible material, a little bit of water expansion due to freezing doesn't damage the collector.

Insulating Pipework

Freezing ceases to be a problem if we can stop the pipes from getting cold in the first place. Whilst there is no sure-fire answer to ensuring that our pipes don't get cold, decent insulation can help reduce the probability of our pipes suffering from freeze-damage. It also makes good energy-efficiency sense, as who wants to go to the trouble of generating solar heat, only to allow it to escape back to the atmosphere?

Antifreeze

A simple solution that necessitates an indirect system is the most commonly used answer to freeze-damage in the UK. It is important to ensure that a non-toxic chemical, which is safe if it enters the foodchain, is used as antifreeze – this ensures that, if there are any leaks in the coil or if the hot-water supply in some way becomes contaminated with water from the solar heating loop, no damage to health can result.

Pump Control

To prevent our solar collector from freezing, we can continuously pump water through it when this condition arises, taking a little bit of heat from the tank and using it to warm the collector, preventing freezing. This system does waste a little energy; however, it can make sense in an environment where freezing is rare.

Trace Heating Cable

It is possible to get a small resistive cable, which can be run around pipework providing a small amount of heat, to stop pipework from freezing in the event of the temperature dropping.

Room to Grow

Our solar installation is heating water that circulates through the solar loop. However, one of the properties of water is that when heated, it expands. If a fault condition were to arise, for example a pump failure, water in a solar collector on a hot, sunny day could reach sufficient temperatures to boil, and the resulting steam would cause undesirable pressure in the system, which could lead to component failure. Our system must be designed so as to allow for a little expansion within the system without damage to any of the system components, property or life. In the system configurations that follow, expansion is dealt with in a number of ways.

In a sealed system, an expansion vessel allows the heat transfer fluid to expand within a sealed loop. As the fluid expands, it compresses the air in the tank and keeps the system at pressure. Contrast this to a vented system, where the expanding transfer fluid is allowed to vent and spill over into a header tank – consequently, the system is not maintained at any pressure.

Circulation

Active Pumped Systems

Mains Powered

Some means is required to keep the working fluid circulating around the solar loop. It is possible to use a pump, powered by the mains – in the UK, this is the most common option. Some installers package the pump with an expansion tank and other necessary hardware for the solar installation into a 'circulation module' permitting easy installation.

Circulation module. (Solar UK)

Circulation Module

Integrally Powered

Some systems, for example, the SolarTwin system, choose not to power their pump from the mains, but instead integrate a photovoltaic cell into their solar collector, which provides power for the circulation pump. The concept is that, as the amount of available solar energy increases, more power is produced by the photovoltaic cell, which increases the pump speed and helps to remove the heat quicker from the solar collector. Whilst this system appears to have an elegant simplicity, it must be noted that the response rates of the electrical system and the thermal system differ; this can result in overheat-

ing when there is insufficient power produced by the PV cell and useful heat not being captured.

Passive Thermosyphoning Systems

Using a pump to circulate the working fluid in a solar water-heating system requires a certain amount of energy in order to power the pump. It is also possible to design a wholly passive system, where a 'thermosyphon' circulates the fluid between the solar collector panel and the cylinder. For a thermosyphon to work, it is important to have the level of the solar collector below that of the tank. The pipework between the collector and tank should consist of smooth bends and straight runs, with the pipework sloping gently down towards the collector. Water that has been heated is less dense than cold water, so, as the collector heats up, the less dense water rises to the cylinder, with denser, colder water flowing down to the collector in its place. Whilst the ideal of a passive system requiring no additional energy can be viewed as a good idea, in reality it is harder to achieve. Making pipes slope gently is difficult; furthermore, there is also an associated risk of freezing in cold weather, which means that this approach to circulation is a rarity in the UK.

Controlling Your Solar Water-Heating System

For effective, safe capture of solar energy, it is necessary to have appropriate controls to ensure that the system operates within its design parameters.

There are some aspects of a solar system we can control, and some we can't: the Sun shining is something that we cannot control; conversely, if a fault arises in the system, generating excess temperature and pressure, we cannot stop the Sun from shining. If we were not aware of a fault developing, this could result in a dangerous situation, with water boiling within the system, and the potential for the system to fail at its weakest point. By careful design, we can control and reduce this risk, with appropriate system controls and safety devices.

For example, taking the system with the simplest solar loop, a thermosyphon system. With no pump to control, because the temperature of water in the cylinder is a function of the amount of sunlight falling on the collector, then with no temperature control the temperatures within the tank can get very high; therefore, control is provided by a thermostatic blending valve, which ensures that the water coming out of the hot taps is at a safe temperature. Systems should also integrate a pressure-release safety valve if sealed, or a vent if open.

Sealed systems should always incorporate pressure-relief safety valves to enable excess pressure to be vented in the event of a fault condition. The valve should be positioned somewhere sensible, where, if hot water is vented, it does not cause any damage. A drip tray or pan should prevent damage to surrounding materials.

For pumped systems, electronic controls can provide a measure of pump control to ensure that the pump is only in operation when necessary. The circuits often employ a comparator to compare two (or more) different temperature sensors, to ensure that the pump is only in operation when the collector temperature is higher than the cylinder temperature. A set point allows the degree of differential between the two sensors to be adjusted. The set point is usually 5–6°C. There may also be some form of user feedback – a display panel showing the temperatures at different parts of the system, or an indicator to show when the pump is operating.

Controls vary in their level of sophistication; whilst a simple differential controller can be built from a handful of components, more sophisticated controllers can run a timed programme of sterilization to kill *Legionella* and other bacteria whose population can increase in warm temperatures, but are killed by high temperatures. More sophisticated variable pump control is also possible, as well as control for supplementary heating.

An example of a solar control can be seen in an example from Solar UK opposite. The pump controller, shown in the bottom-left of the picture, is what performs the control functions. The display panel, in the top left-hand side corner,

Solar controller.
(Solar UK)

provides feedback to the user in the form of a digital display, and a light indicating when the tank is hot and when the pump is in operation. The pump under control is shown in the bottom-left corner; this connects to the controller.

In the top right-hand corner, we can see the three temperature sensors, which are used to take measurements at different positions in the system. For accurate control, it is essential that these sensors make good thermal contact with the surfaces that they are monitoring. Your product supplier should advise you on the correct method of installation for their particular sensors. However, it should be noted that there are a variety of products on the market to aid thermal conduction; for example, pastes and adhesives, which ensure a good thermal bond.

For systems such as the 'SolarTwin' system, no sophisticated controls are required, with the pump speed being controlled by the amount of insolation on the solar panel adjacent to the collector.

Solar Water-Heating System Configurations

Pumped, Indirect, Open and Vented System

This system employs a separate pumped loop for the solar circuit; being open, a vent is provided to allow for expansion. This vents into a header tank above the level of the collector to keep the system at a working pressure. Because the system is open, no safety pressure-relief valves are required. However, a drain cock is provided to drain down the system.

Pumped, Indirect, Sealed and Fully Filled System

The pumped indirect system does not have a vent; unlike the open system, it is sealed. A pressure gauge provides some monitoring of the system pressure, and a safety valve is provided to allow expansion beyond the design constraints of the system to be vented. An expansion vessel allows for some variation of pressure within the system. An air vent needs to be provided at the top of the system, to allow any air to be bled out of the system. It is a pumped system.

Pumped, Indirect, Sealed and Drainback System (Vessel on Return)

We can make an open system with a drainback vessel on either the flow or return. In this system,

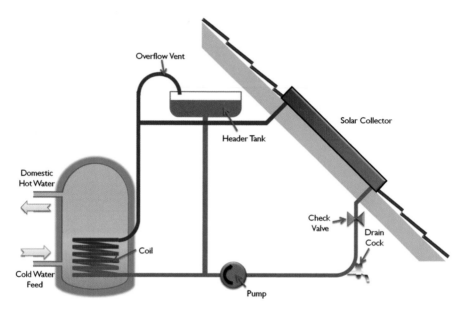

ABOVE **Pumped, indirect, open vented system.**

RIGHT **Pumped, indirect, sealed and fully filled system.**

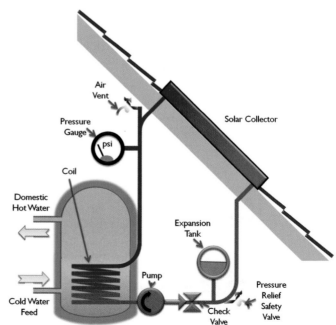

the pump is above the drainback vessel, sucking water out the vessel when the system resumes operation. When the pump stops, the water above the pump drains out of the collector, back into the drainback vessel, which is sized to accommodate the water in the system above that point.

Pumped, Indirect, Sealed and Drainback System (Vessel on Flow)

By contrast, a drainback system with the vessel on flow has the vessel on the opposing side of the system to the pump. When the pump stops, the water finds its level, draining back through

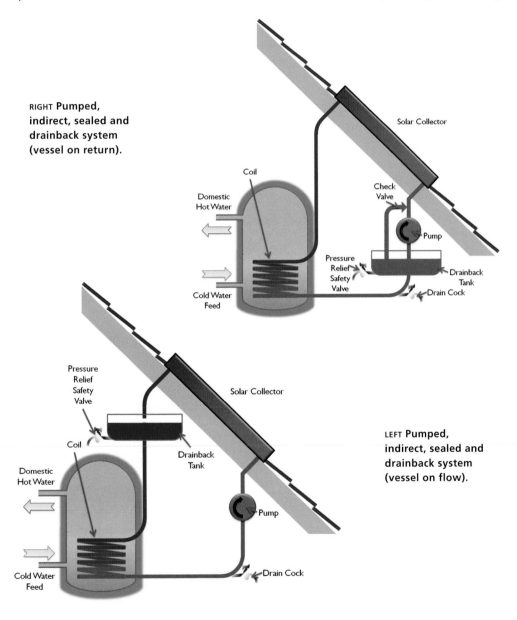

RIGHT **Pumped, indirect, sealed and drainback system (vessel on return).**

LEFT **Pumped, indirect, sealed and drainback system (vessel on flow).**

the pipework, where it is held in the drainback tank until the pump resumes operation.

Thermosyphoning, Indirect, Open Vented System

In a thermosyphoning system, no pump is required to circulate the water around the system, instead the natural current created by heated water being less dense than cold water causes the fluid to circulate round the system. Thermosyphoning systems can be either indirect, with a flow of water through a coil, heating water in the cylinder, or direct, where the water flowing through the panel is the water that will eventually be used in the hot-water outlets.

In this system, the elements of the open direct system are integrated with a thermosyphoning loop circulating water through the solar circuit. It is important to note that careful design of pipework is absolutely critical to ensuring that the system syphons correctly. The nature of the system necessitates that the collector is below the level of the tank, and the open, vented nature of the system, requires that there is a header tank above the level of the cylinder. Because of this configuration, mounting the collector on the roof will nearly always be impossible. This system is more appropriate for a collector mounted low down, say at the base of a wall, with a clear view of the sky and no obstructions.

Thermosyphoning, Direct, Open Vented System

In a thermosyphoning direct system, the tank is again mounted below the collector, with a

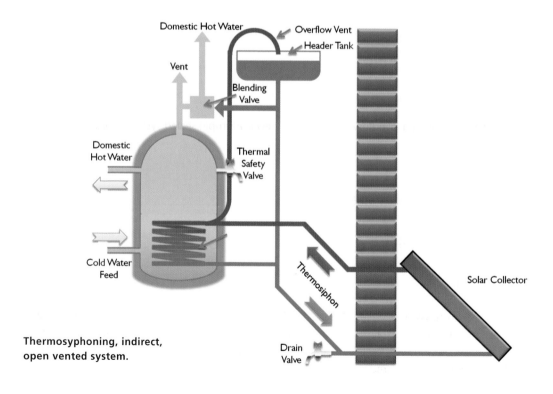

Thermosyphoning, indirect, open vented system.

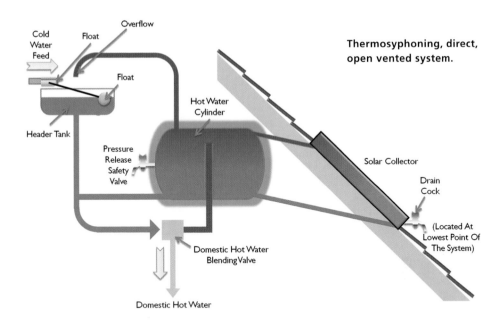

Cold Water Feed

Float

Overflow

Float

Header Tank

Hot Water Cylinder

Pressure Release Safety Valve

Thermosyphoning, direct, open vented system.

Solar Collector

Drain Cock

(Located At Lowest Point Of The System)

Domestic Hot Water Blending Valve

Domestic Hot Water

shallow drop down to the solar collector. The thermosyphon takes care of the circulation of water back to the tank and, because there is not the complexity of a coil, the water in the tank is the water that will appear at the taps. In an open vented system, provision will still need to be made for a header tank to provide pressure to the cylinder, and also an overflow to allow for expansion.

Retrofit Solar Water-Heating System Configurations

Retrofit Heat Exchanger

Retrofit heat exchangers were discussed earlier in this chapter. They fit into an existing hot-water cylinder in the place of the immersion heater, which is usually fitted. They benefit from being relatively simple to install and they do not necessitate a new cylinder; however, they are a less efficient solution. Being mounted at the top

of the tank, they tend to heat the warm water – in contrast to a specially designed solar cylinder, which will heat the water at the bottom of the cylinder, which is coldest, and where more efficient heat-transfer can take place. It is of course possible to have a number of variations in the configuration of the solar circulation circuit; however, a simple sealed system is often employed.

SolarTwin System

The SolarTwin system has been designed to be retrofitted onto an existing heating system. The controls for the system are relatively simple, with a pump being powered directly from a small solar panel integrated directly with the hot-water cylinder. As the amount of sunlight falling on the solar panel increases, the pump speeds up, which should correspond with an increased rate of heating of the solar collector. This system does have its critics, but it has proven a popular solution in the UK.

107

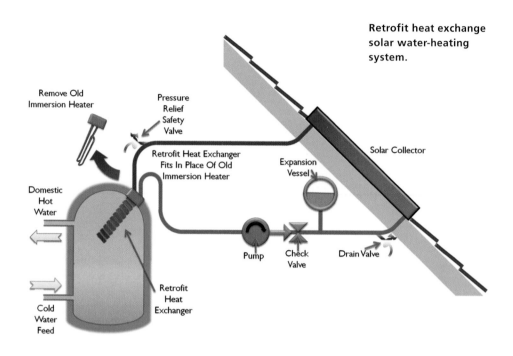

Retrofit heat exchange solar water-heating system.

Remove Old Immersion Heater

Pressure Relief Safety Valve

Retrofit Heat Exchanger Fits In Place Of Old Immersion Heater

Solar Collector

Expansion Vessel

Domestic Hot Water

Pump

Check Valve

Drain Valve

Retrofit Heat Exchanger

Cold Water Feed

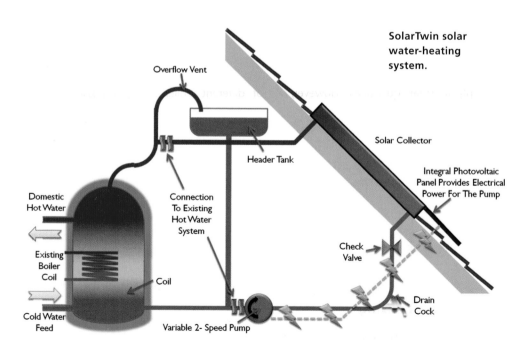

SolarTwin solar water-heating system.

Overflow Vent

Solar Collector

Header Tank

Integral Photovoltaic Panel Provides Electrical Power For The Pump

Domestic Hot Water

Connection To Existing Hot Water System

Check Valve

Existing Boiler Coil

Coil

Drain Cock

Cold Water Feed

Variable 2- Speed Pump

Legionella Survival Characteristics Over a Range of Temperatures	
Temperature	*Legionella* **Survival Characteristics**
70–80°C	Disinfection range – *Legionella* die completely.
At 66°C	*Legionella* die within 2 minutes of exposure to temperature.
At 60°C	*Legionella* die within 32 minutes of exposure to temperature.
At 55°C	*Legionella* die within 5–6 hours of exposure to temperature.
50–55°C	*Legionella* can survive but do not multiply in this range of temperature.
20–50°C	*Legionella* can multiply and grow within this range of temperature.
35–46°C	Ideal growth range for *Legionella.*
Below 20°C	*Legionella* can survive but are dormant below this temperature.

Solar Water-Heating Systems and *Legionella*

Any domestic hot-water system, whether conventional or solar, can serve as a breeding ground for *Legionella* bacteria, which can cause infections classed as legionellosis, resulting in either legionnaire's disease or pontiac fever. It should be noted that less than 5 per cent of the population are susceptible to infection by *Legionella* but, if infected, the consequences can be serious.

Legionella bacteria are present in the domestic water supply in small quantities. However, if water lays dormant, the bacteria have the chance to rapidly multiply and spread. The above table shows the characteristics of *Legionella* bacteria over a range of temperatures.

For this reason, it is essential that, if there is insufficient insolation to heat the water to the desired temperature, supplementary heating is provided, to raise the temperature of the water to a level which will kill *Legionella* bacteria.

Integrating Solar Heating with Your Existing Plumbing

There is often a hard decision to be made when trying to integrate a solar hot-water system with an existing system: whether to retain elements of the existing system, or to start afresh with the installation.

Some older combination boilers can react unexpectedly to pre-heated water from a solar installation, whilst more modern combi-boilers, which are sold as being 'solar compatible', have more sophisticated controls and can deal with the influx of hot water. There is a serious risk when using any boiler not designed to work with a pre-heated input, but because of the vast variety of models and performance of different boilers, it is impossible to offer generic advice on this matter other than to say that it is neccessary to consult a qualified heating engineer who is familiar with your existing boiler and solar water heating systems, when adding solar pre-heat to an existing installation.

If retrofitting solar to an existing boiler installation, do not do so without first consulting the manufacturer's documentation – and if this does not yield sufficient information, ask the manufacturer, and do not have any work carried out until you have had a thorough appraisal from a competent plumber who specializes in heating.

CHAPTER 6

SOLAR HEATING FOR SWIMMING POOLS

In the long run, extra money invested in solar heating for pools will pay back in energy savings.

- Solar pool-heating has very few moving parts and so there is very little to maintain or go wrong.
- Safety concerns with a gas or electric heater installation are eliminated.
- If you try to sell your home, there is a measure of reassurance for prospective buyers that they will not be faced with unduly high costs for running a pool.
- Swimming pools are a luxury purchase – not a necessity. They consume massive amounts of energy, which traditionally produces carbon emissions. By heating your pool with solar energy, you are being pro-active in reducing your carbon emissions, reducing your environmental impact and can enjoy guilt-free swimming in the knowledge that the energy used to heat your pool is not damaging the environment.

Demand and Availability of Solar Energy

Unlike space heating for the home, the period for heating demand for a swimming pool and the availability of solar energy coincide well. Indeed, for outdoor pools, it is on clear sunny days, which are favourable for solar swimming-

pool heating, that use of a pool is most likely. For an indoor pool, which is used all-year round, you will need an auxiliary pool heater.

Swimming-Pool Heat Losses

Your swimming pool is fighting a losing battle to retain its heat energy. For comfortable swimming, the temperature of your pool needs to be warmer than its surroundings – however, this creates a heat differential with the warmth of the pool wanting to escape into the cold environment. Maintaining such a large body of water at a snug temperature requires an enormous amount of energy – we need to find out where this energy is being wasted.

Evaporative Losses

Evaporation accounts for the bulk of heat lost from a swimming pool. On a warm summer day, 50mm of water is lost from swimming pools every week as the water evaporates. The process takes with it not only the water from the pool (and the heat with it) but also the chemicals used to keep the pool's water clean and clear (which adds to the cost of running a pool). These losses are affected to some degree by the environmental factors – a warm atmosphere will receive more evaporated water than a humid one – however, a pool

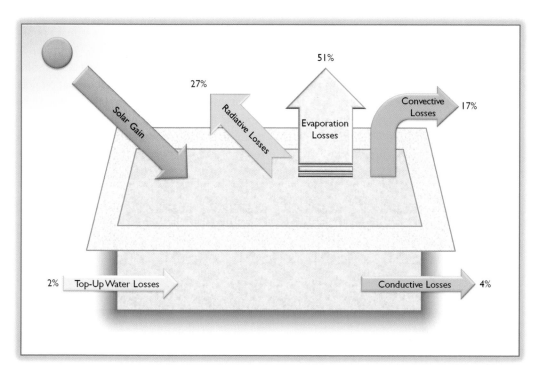

Representative thermal losses from an average swimming pool.

cover can help to reduce the effects of evaporative losses.

Radiative Losses

Radiation is a process by which heat is lost, by transmission through air from a warm body to a cold one. Radiation can account for anywhere between 25 and 35 per cent of heat loss from pools.

Convective Losses

When the pool pump is operating, the warm and the cool water in the pool are constantly being mixed. However, when the pump stops operating, the less dense, warm water will rise to the top of the pool, whilst the dense, cool water will sink to the bottom. By this process the heat embodied in the water moves to the top of the pool, where it comes into contact with the cold air. The heat loss will increase if wind is blowing on the surface of the water. Convective losses account for between 15 and 25 per cent of heat loss from swimming pools. However, by using a pool cover, the warm layer of water – normally in contact with the air – can be insulated and the heat retained.

Conductive Losses

The walls and floor of your swimming pool are in contact with the surrounding earth or, if your swimming pool is raised from the ground, your walls will be supported by a raised masonry (or similar) wall. The water in contact with these surfaces will transfer heat as a result of the physical contact and temperature differential. Conductive losses account for around 4 per cent of the heat lost from your swimming pool. This figure may be greater for an 'out of ground' pool, or may be increased by the conditions of the ground in which your pool is sited, for example a high water-table will tend to conduct heat away from the earth surrounding the pool.

Water Losses

Invariably, in the course of using a swimming pool, entering and exiting it, diving in or bombing, water will be lost from the swimming pool – by splashing out of the pool or by dripping off your body as you walk out of the pool. Some water will also be lost as a result of evaporation. All this water that escapes from the pool must be replaced in order to ensure that the water level does not drop below the skimmer line. Invariably the replacement water is going to come from the cold main's water supply. This water will reduce the temperature of the pool if it enters unheated, and therefore heat energy is required to replace the heat lost by the escaped water.

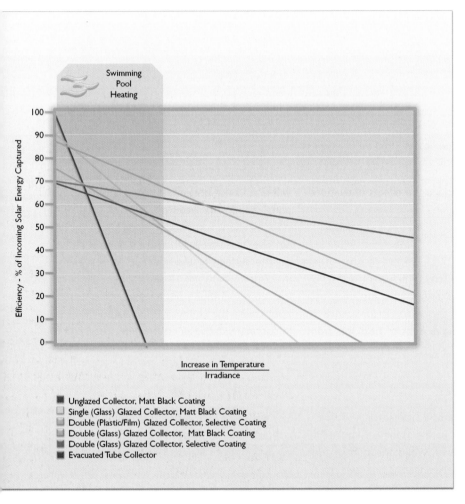

Band in which swimming-pool collectors operate.

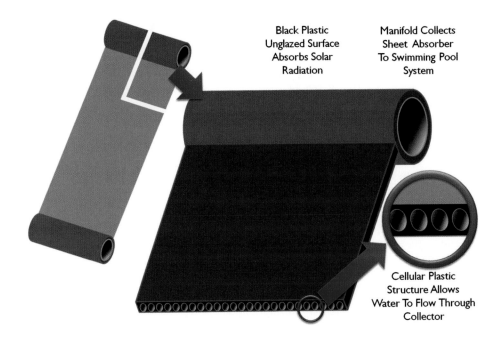

Black Plastic
Unglazed Surface
Absorbs Solar
Radiation

Manifold Collects
Sheet Absorber
To Swimming Pool
System

Cellular Plastic
Structure Allows
Water To Flow Through
Collector

Flooded plate collector construction.

Solar Collectors for Pool Heating

Swimming pools present a different scenario to heating water for domestic hot water. The temperature difference between the water to be heated and the target temperature is generally lower with swimming-pool installations, which means that we can use less sophisticated solar collectors for swimming-pool heating.

The temperature to which we aim to heat our swimming pool (around 28–34°C) is essentially much lower than the requirements for domestic hot water, presenting a less challenging situation. We can see, looking at the graph opposite, that for the band in which swimming-pool collectors operate, an unglazed collector can provide good performance.

When selecting a suitable collector, if direct heating of the swimming pool water is to be employed, one of the considerations is that the corrosive chemicals required to keep swimming pools clean, clear and sanitary, present a challenge – unglazed plastic collectors will not corrode with swimming-pool water flowing directly through them.

Unglazed Collectors

Unglazed collectors are generally manufactured from a plastic, usually polypropylene or rubber, impregnated with carbon black. It is the carbon black that gives the collectors their colour and allows them to absorb the Sun's solar radiation. However, the UV component of sunlight can cause some plastics to degrade, so a UV inhibitor is included to ensure that the panels enjoy a long life and do not fail prematurely.

With an unglazed collector, it is essential to ensure that the pipes do not remain filled with water in the event of temperatures dropping

113

below freezing, as ice may form, expanding in the collector and causing damage. For this reason, unglazed collectors are generally designed with a shallow angle to the array, which permits them to 'drain back' in the event of cold weather.

Unglazed solar collectors used for swimming-pool heating are generally known as 'flooded plate' collectors; they consist of a hollow cellular plastic, with channels running vertically. The swimming-pool water flows through these channels that make up the collector area, with the collector effectively being 'flooded' with water. At either end of the sheet, a manifold connects the pipework of the swimming-pool heating system to the collector area.

Another approach to this problem is to use evacuated tubes, an approach favoured by Solar UK. Their LaZer tracker system consists of eight evacuated tube panels, mounted on a solar tracker, which follows the Sun. This is a more sophisticated system than simple unglazed collectors, and by tracking the Sun, more solar energy can be captured than with a static installation.

Collector Area

Many of the rough and ready estimations of collector areas are based on the surface area of your pool. If you are using simple, unglazed collectors, a ready reckoner is to aim to have around 50–100 per cent of the area of your pool in the collector area. The larger you size your collector area, the longer the swimming season for your pool will be.

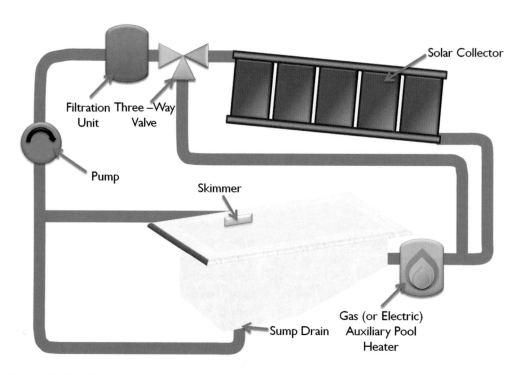

Schematic of swimming-pool heating system.

Evacuated solar panels on a pool house. (Solar UK)

Indications for a Larger Collector Area

There are a number of indications for using a larger collector area than the ready-reckoner would seem to suggest.

▪ More than a quarter of the collector array will be unduly shaded.
▪ You prefer a particularly warm pool.
▪ You live on a particularly windy site.
▪ Site/planning conditions do not permit you to face the collectors due south.

Swimming-Pool System Configuration

Swimming-pool heating systems have a relatively simple system configuration that is easy to understand. It is a relatively simple task to convert an existing swimming-pool installation to function with a solar collector. Your existing system will likely comprise of a pump, which feeds a filtration unit, which then passes water to a gas or electric water heater for return to the pool.

With a solar water-heating system, we insert a diverter valve between the filtration unit and the water heater, allowing us to divert the swimming-pool water through the solar collector, before then feeding it through the gas or electric heater. When the temperature of the collector area is higher than the temperature of the water, we can divert the water to first flow around the solar loop and harness the solar energy. However, when it is cold, we simply send the water straight to the 'conventional' heater. This arrangement allows us to retain the 'conventional' heater for supplementary heat when required, or to switch it off when it is unnecessary.

Installation Options

Where and how you install your panels will largely depend on site conditions and the available area for mounting panels.

Pool-House Roof

One option is to build (or utilize an existing) a pool house adjacent to the pool. A pool house gives you space to get changed, keep the swimming-pool plant and maybe even a room for a (solar) shower. It also provides a convenient roof area for mounting solar panels. There is an advantage to mounting panels on a pool-house roof, in that they are elevated by a number of metres – raising them above low-level obstructions like small trees, fences and other things in the pool-side area, which could shade the

Evacuated tube solar panel tracker for swimming pool. (Solar UK)

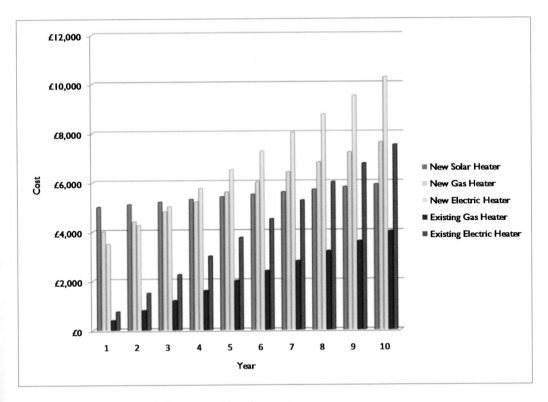

Example costings for five different pool heating options.

solar collector. It goes without saying that it is necessary to ensure that the structure of the pool house is sufficient to support the weight of the solar collectors and associated fixtures and fittings.

Ground Adjacent to the Pool

One option, if you have plenty of space, is to mount the collectors at ground level adjacent to the pool. However, be aware when mounting collectors at low level of physical obstructions, such as planting, sunshades and other objects in the garden that could overshadow the panels. Mounting at high level reduces the chances of some of these problems occuring.

Solar Tracker

If you are using more expensive evacuated tubes to heat your swimming pool, you could employ a solar tracker that follows the Sun's path. This requires a clear area of ground for installation, with a good view of the sky.

Evaluating the Economics of Different Pool-Heating Options

The graph above examines five different options for a swimming pool – the shape, size and heating requirements of swimming pools differ vastly, so this serves as a discussion point for your own installation rather than relying on the figures. We are working on the following premises, which at the time of publication can generally be thought of to hold true; however, with volatile energy markets and the prices of products changing day by day it would be impossible to state definitive values for energy prices and products. Instead, please consider the graph as indicative of an example swimming pool, and base your own calculations on the energy prices and cost of products at the time you price your installation.

A solar installation of equivalent output will generally cost more than conventional pool heaters. Gas pool-heaters will also tend to be more expensive than an equivalently rated electric model; however, it must also be considered that gas is cheaper per unit than electricity.

We assume that the fixed costs of an existing installation have already been met, and that all that is required is the energy to operate them. We have not allowed for maintenance in this calculation – however, it must be borne in mind that an existing installation is likely to be more costly to maintain than a brand-new one, which may be covered by a guarantee warranty and is not likely to need serious repair or attention for some years.

The new solar installation is a higher fixed cost than the other installations to begin with; however, year on year, they do not require the energy input of a conventional pool-heater. Thus the variable costs of energy are removed from the equation – of course, you can couple a solar installation with a conventional heater, allowing you to use free energy in the months with better conditions for solar heating, but supplementing this with paid-for energy to prolong the swimming season.

The other option to compare against is an unheated pool. However, this severely restricts the number of months of the year for which you can enjoy your pool – installing solar heating will greatly prolong the swimming season.

Many suppliers of swimming-pool heating systems estimate that it will take around five years for a swimming-pool heating system to pay back, because of the vast amount of conventional energy used to heat swimming pools.

CHAPTER 7

COSTING YOUR SOLAR INSTALLATION

The attitude of many in the UK seems to favour spending now, paying later; in making paying loan repayments that, while cheap in the short term, add up to vastly more than the cost of the item over the long run. This is analogous to the way that we view energy – we are prepared to pay for cheap energy from fossil fuels and nuclear power which, though economical in the short term, does not take account of the 'cost' incurred (financial and environmental) in the long run. As responsible consumers of energy we must evaluate the true value of renewable energy by looking at the cost – not only to ourselves, but also to the environment. The short-term attitudes we seem to have adopted are not helpful when looking at renewable energy installations, as they require the foresight to think for the long term, to invest wisely today in order to reap dividends for the future.

There are three spheres of sustainability: environmental, economic and social. For a project to be truly sustainable, it must be viable in all these three spheres of existence, being a solution that is environmentally sensitive, financially sound and also a solution that is socially sustainable.

Any investment in exploiting solar energy in your home is going to involve a large expenditure in the short term; however, the reward will pay back slowly over a long time. You'll see the repayment manifested in lower energy bills

(financial), the feeling that your household energy use is having a far-reduced impact on the environment (environmental) and, eventually, as an early adopter, you might see the social impact that other people on your street follow your example and have solar devices installed on their property too (social) – the reward is thus threefold.

In order to quantify the benefits that a solar installation can bring to us, we must project the cost of the project over its life-cycle and work out at what rate the benefit that the installation gives us will accrue. We can place a cost on that benefit by, for example, looking at the amount it would cost us to provide that benefit by other methods. We can then reconcile these figures to tell us how long the installation will take to 'pay for itself', based upon the costs that we avoid paying for deriving that benefit from other means.

What Will My Installation Cost?

We live in a volatile market of sustainable technology. The massive investments in research and development are helping to drive the cost of sustainable technology down significantly. The price of PV modules has plummeted – and new technologies based on nanotechnology and enhanced materials science look to make the cost of photovoltaic cells fall even further.

Meanwhile, looking at the prospects for heating, more and more manufacturers are entering the marketplace.

The price of photovoltaic panels has fallen over 500 per cent in the past two decades. The cost of a photovoltaic installation is currently around £6,000 to £7,000 per kWp (kilowatt peak), which means that for an average domestic installation providing 2kWp, an installation will cost in the region of £12,000 to £14,000 – however, this is before any grant funding that may be available. (Kilowatt peak, abbreviated kWp, represents the output of a solar module under ideal standard test conditions.) There are many moves to reduce the cost of photovoltaic cells – one US company 'Nanosolar' has set itself the target of delivering thin-film solar cells for 99 cents (50 pence) a Watt, which equates to £500 per kWp – a significant saving. However, it is only through the continued investment of early-adopters and consumers now that these technologies will reach the marketplace.

Bear in mind also, that there are factors that will make the total cost of installation more or less – if you opt for a larger installation, there may be economies of scale (a larger inverter may be cheaper per 'kWp'), cables only need to be run once, some hardware will only be required once for each installation and so on. Conversely, if you have specific requirements for a certain type of photovoltaic cell or panel, this may push the price of an installation up; for example, if your installation has any particularly awkward features, such as a tricky roof shape.

Life-Cycle Cost Analysis

The life-cycle costs take into account the cost that an installation will incur over its operating lifetime. Any investment is going to require some supplement in order to maintain the value of that initial capital investment. If you don't paint your woodwork, it will soon become rotten – and so, in the same way, to ensure correct function, your solar installation may require some planned maintenance. The following simple equation, which is explained below, can be used to help you calculate the life-cycle costs of your installation:

$$LCC = (C - G) + (M \times n) + (E \times n) + R - S.$$

LCC = life-cycle costs.

C = capital costs, which are the initial costs of the installation, including all equipment costs and the labour for installing that equipment.

G = grants awarded, which is the value of any grants that are awarded to subsidize the capital costs of installation – you can read about these later in this book.

M = maintenance costs, which is the annual cost of all planned maintenance engineering that needs to take place on an installation. It could include such tasks as replacing bushes, seals or bearings, but will not include the cost of planned replacements or energy input to a system.

n = number of years.

E = energy costs, which are the annual costs of the 'input' of energy required to run a system. For example a solar photovoltaic system requires no energy to run – it produces all of its own energy, so the value here will be '£0'; however, a solar water-heater may require an input of energy to run a water pump (unless it's a Solartwin system) – here there will be an annual amount of energy consumption, which needs to be added to the amount the installation will cost over its life-cycle.

R = replacement costs, which are the costs of any items of equipment that are expected to fail and need replacement over the lifetime of the installation. It may include items such as fuses, which are expected to blow in the event of a fault, or it could include items with a shorter service-lifetime than the life of the whole installation.

S = salvage value, which is the final value of all assets in the last year of the life-cycle period that can be removed and sold on for re-use, spares, scrap or recycling. Commonly a figure of 20 per cent of the original cost of components only (not including labour and installation) is assigned.

Annuitising the Cost of Borrowing

'A bird in the hand, is worth two in the bush', in other words, money 'today' will always be worth more than money 'tomorrow'; given the chance to be paid in advance or arrears, advance wins hands down. Given the preference, many people will always opt to choose benefits today and pay for them later. With solar energy systems, there is a 'capital' cost involved – we are making an investment in a piece of equipment that will have a usable life of over a year and we will continue to derive benefit from that system for several years to come. We need to make an objective comparison between the benefits that we derive from the solar installation and the opportunity lost by having made that investment and not having that money available for other uses.

People will invest their money in schemes and projects, but expect to be compensated for not being able to have their money available – and as a result, will seek 'interest' on the amount loaned. The rate of interest will depend on lots of factors, market conditions, how risky the lending is considered, how quickly the person could recoup their full amount if they needed access to it.

There are lots of mechanisms through which you could borrow money to pay for a solar installation: you could borrow against the value of your home, you could take out a personal loan, you could even stick some of it on a credit card – all these different methods of borrowing will command different interest rates.

We need to take the cost of our installation, work out the percentage interest we will have to pay on the money borrowed to finance the installation, and then work out the equivalent annual payment for the system, or the annuitised cost. We can then work out how much we anticipate our installation to save every year in energy bills, how many years the installation is expected to last for, to get some gauge of how long the system will take to 'pay back' its installation cost.

The tables on page 123 give us an indication of the annuitised cost of a sum of £1,000 over various discount rates, and over a number of years. The table on page 124 shows us what the cumulative cost of the borrowing over the term is. Some shading has been added to help you make comparisons between different repayment regimes.

Grant Funding – Low Carbon Buildings Programme

Grant funding can help make the economics of a renewable energy installation seem more attractive. However, the present state of grant funding in the UK has been severely criticized for its disjointed nature, and 'lack of funds in the pot'. Grant funding in the UK is notoriously hard to obtain, with stories widely reported in the mass media of the total sum of grant money 'running out' within hours of being released, and frantic homeowners wanting to green their home, frantically trying to submit applications for funding only to find they have been pipped to the post only minutes before. The government-provided grants for renewable energy development have gone through several iterations and titles, however, for domestic installations, grant funding currently falls under the 'Low Carbon Buildings Programme'. Grants of up to £2,500 are available per property towards the cost of a certified installation installed by an approved installer.

BERR supply a list of certified installers whose installations are eligible for funding at the following web address:
www.lowcarbonbuildings.org.uk/info/installers/
Furthermore, information on how to apply for a grant is available here:
www.lowcarbonbuildings.org.uk/how/

Annuitised Cost

Real Discount Rate %

Capital Repayment Period Years	0%	1%	2%	3%	4%	5%	8%	10%	13%	15%
1	£1,000.00	£1,010.00	£1,020.00	£1,030.00	£1,040.00	£1,050.00	£1,075.00	£1,100.00	£1,125.00	£1,150.00
2	£500.00	£507.51	£515.05	£522.61	£530.20	£537.80	£556.93	£576.19	£595.59	£615.12
3	£333.33	£340.02	£346.75	£353.53	£360.35	£367.21	£384.54	£402.11	£419.93	£437.98
4	£250.00	£256.28	£262.62	£269.03	£275.49	£282.01	£298.57	£315.47	£332.71	£350.27
5	£200.00	£206.04	£212.16	£218.35	£224.63	£230.97	£247.16	£263.80	£280.85	£298.32
6	£166.67	£172.55	£178.53	£184.60	£190.756	£197.02	£213.04	£229.61	£246.68	£264.24
7	£142.86	£148.63	£154.51	£160.51	£166.61	£172.82	£188.80	£205.41	£222.60	£240.36
8	£125.00	£130.69	£136.51	£142.46	£148.53	£154.72	£170.73	£187.44	£204.83	£222.85
9	£111.11	£116.74	£122.52	£128.43	£134.49	£140.69	£156.77	£173.64	£191.26	£209.57
10	£100.00	£105.58	£111.33	£117.23	£123.29	£129.50	£145.69	£162.75	£180.62	£199.25
15	£66.67	£72.12	£77.83	£83.77	£89.94	£96.34	£113.29	£131.47	£150.76	£171.02
20	£50.00	£55.42	£61.16	£67.22	£73.58	£80.24	£98.09	£117.46	£138.10	£159.76
25	£40.00	£45.41	£51.22	£57.43	£64.01	£70.95	£89.71	£110.17	£131.94	£154.70
30	£33.33	£38.75	£44.65	£51.02	£57.83	£65.05	£84.67	£106.08	£128.76	£152.30
35	£28.57	£34.00	£40.00	£46.54	£53.58	£61.07	£81.48	£103.69	£127.06	£151.13
40	£25.00	£30.46	£36.56	£43.26	£50.52	£58.28	£79.40	£102.26	£126.13	£150.56
45	£22.22	£27.71	£33.91	£40.79	£48.26	£56.26	£78.01	£101.39	£125.63	£150.28
50	£20.00	£25.51	£31.82	£38.87	£46.55	£54.78	£77.07	£100.86	£125.35	£150.14
75	£13.33	£19.02	£25.86	£33.67	£42.23	£51.32	£75.33	£100.08	£125.02	£150.00
100	£10.00	£15.87	£23.20	£31.65	£40.81	£50.38	£75.05	£100.01	£125.00	£150.00

Total Repayment

Real Discount Rate %

Capital Repayment Period Years	0%	1%	2%	3%	4%	5%	8%	10%	13%	15%
1	£1,000.00	£1,010.00	£1,020.00	£1,030.00	£1,040.00	£1,050.00	£1,080.00	£1,100.00	£1,113.00	£1,150.00
2	£1,000.00	£1,015.02	£1,303.10	£1,045.22	£1,060.39	£1,075.61	£1,113.86	£1,152.38	£1,191.18	£1,230.23
3	£1,000.00	£1,020.07	£1,040.26	£1,060.59	£1,081.05	£1,101.63	£1,153.61	£1,206.34	£1,259.79	£1,313.93
4	£1,000.00	£1,025.12	£1,050.50	£1,076.11	£1,101.96	£1,128.05	£1,194.27	£1,261.88	£1,330.83	£1,401.06
5	£1,000.00	£1,030.20	£1,060.79	£1,091.77	£1,123.14	£1,154.87	£1,235.82	£1,318.99	£1,404.27	£1,491.58
6	£1,000.00	£1,035.29	£1,071.15	£1,107.59	£1,144.57	£1,182.10	£1,278.27	£1,377.64	£1,480.08	£1,585.42
7	£1,000.00	£1,040.40	£1,081.58	£1,123.54	£1,166.27	£1,209.74	£1,321.60	£1,437.84	£1,558.22	£1,682.52
8	£1,000.00	£1,045.52	£1,092.08	£1,139.65	£1,188.22	£1,237.77	£1,365.82	£1,499.55	£1,638.66	£1,782.80
9	£1,000.00	£1,050.66	£1,102.64	£3,155.90	£1,210.44	£1,266.21	£1,410.90	£1,562.76	£1,721.34	£1,886.17
10	£1,000.00	£1,055.82	£1,113.27	£1,172.31	£1,232.91	£1,295.05	£1,456.86	£1,627.45	£1,806.22	£1,992.52
15	£1,000.00	£1,081.86	£1,167.38	£1,256.50	£1,349.12	£1,445.13	£1,699.31	£1,972.11	£2,261.46	£2,565.26
20	£1,000.00	£1,108.31	£1,233.13	£1,344.31	£1,471.64	£1,604.85	£1,961.84	£2,349.19	£2,761.91	£3,195.23
25	£1,000.00	£1,135.17	£1,280.1	£1,435.70	£1,600.30	£1,773.81	£2,242.77	£2,754.20	£3,298.59	£3,867.49
30	£1,000.00	£1,162.44	£1,339.50	£1,530.58	£1,734.90	£1,951.54	£2,540.14	£3,182.38	£3,862.80	£4,569.01
35	£1,000.00	£1,190.13	£1,400.08	£1,628.88	£1,875.21	£2,137.51	£2,851.90	£3,629.14	£4,447.07	£5,289.72
40	£1,000.00	£1,218.22	£1,462.23	£1,730.50	£2,020.94	£2,331.13	£3,176.01	£4,090.38	£5,045.37	£6,022.48
45	£1,000.00	£1,246.73	£1,525.93	£1,835.33	£2,171.81	£2,531.78	£3,510.52	£4,562.60	£5,653.21	£6,762.55
50	£1,000.00	£1,275.64	£1,519.16	£1,943.27	£2,327.51	£2,738.84	£3,853.62	£5,042.96	£6,267.36	£7,506.93
75	£1,000.00	£1,426.21	£1,939.13	£2,525.10	£3,167.18	£3,849.12	£5,649.91	£7,505.90	£9,376.37	£11,250.32
100	£1,000.00	£1,586.57	£2,320.27	£3,164.67	£4,080.80	£5,038.31	£7,505.43	£10,000.73	£12,500.10	£15,000.01

CHAPTER 8

RELEVANT LEGISLATION

Disclaimer: Legislation is constantly changing and dynamic – this chapter represents an introduction to the homeowner of the relevant legislation concerning a solar installation. However, judgement on a particular installation may be peculiar to the locality of the installation and the specific factors concerning its implementation, so do not proceed without first clearing the project with your local planning authority.

Planning Permission

A common query of homeowners wishing to install a solar installation is 'Will I require planning permission/building consent?'. The answer requires some consideration, and it is always worth a courtesy phone-call to your local planning department to ensure that they are 'in the loop' and you are on the 'right side of the law'. The chances are, if you are well-armed with the facts and know what you can and can't do, you are less likely to encounter any problems.

Under planning law, not all works necessarily require 'planning permission'. There are some works that fall under the category of a 'permitted development', which means that development can proceed with the minimum of fuss and red-tape.

Planning Permission Checklist

Were the 'permitted development rights' of your property removed for alterations to the roofline?

Yes ☐ No ☐

Is your house covered by an 'article 4' declaration and does the proposed installation front an open space or highway?

Yes ☐ No ☐

Would the proposed installation of solar panels be at a different angle to the roofline?

Yes ☐ No ☐

Do you live in a flat?

Yes ☐ No ☐

Is your property a listed building?*

Yes ☐ No ☐

Are your panels of an unusual or non-standard design?

Yes ☐ No ☐

Is your property within the boundary of a 'conservation area'?**

Yes ☐ No ☐

*If 'Yes', listed building consent is also required.
**If 'Yes', conservation area consent is also required.

You can find out more about Planning in the UK and details for your local planning department at www.planningportal.gov.uk

Building Regulations Checklist

If you are planning to 're-roof' your property, and include roof-integrated photovoltaics, you will need to seek building regulations approval.

Part L of the Building Regulations

Part L of the Building Regulations 2000 'Conservation of Fuel and Power' came into force on 1 April 2002.

Part L introduced in 2006 encourages the use of low and zero-carbon energy generation technologies. However, it should be noted that Part L puts a limit on solar gain.

Listed Building Consent

Any building that is listed requires authorization to be obtained before any works are undertaken for alteration, extension, or any modificatiosn which could substantially affect the character of the property. Furthermore, any buildings or structures which are considered within the 'curtilage' or grounds of a listed building will also require consent. Any alteration to a listed building without consent is a criminal offence, so consult with your local authority's planning department before considering a solar installation. Planners may be more amenable to freestanding systems that do not directly affect the structure and are independent from it, so give this consideration as a way of proceeding.

Conservation Area Consent

The Government has issued the following guidance on developments in conservation areas, it reads:

> In sites with nationally recognised designations (Sites of Special Scientific Interest, National Nature Reserves, National Parks, Areas of Outstanding Natural Beauty, Heritage Coasts, Scheduled Monuments, Conservation Areas, Listed Buildings, Registered Historic Battlefields and Registered Parks and Gardens) planning permission for renewable energy projects should only be granted where it can be demonstrated that the objectives of designation of the area will not be compromised by the development, and any significant adverse effects on the qualities for which the area has been designated are clearly outweighed by the environmental, social and economic benefits.

Party Wall Act

If you are building within 6m of a shared boundary, you may need to consult with your neighbour under the 'Party Wall Act 1996'.

INDUSTRY BODIES, STANDARDS AND MARKS OF QUALITY

Reassurances of Quality

For any product or service, it pays to know the industry bodies, standards, certification and marks of quality that are used within that industry to signify good workmanship and a competent standard of service.

The Clear Skies Accreditation Scheme has been superceded by the Microgeneration Certification Scheme (*see* below).

Low Carbon Buildings Programme Accreditation

The Low Carbon Buildings Programme is administered by the Department for Business, Enterprise and Regulatory Reform. It supercedes previous programmes administered under the DBERR's former guise of the Department of Trade and Industry. DBERR manages the Low Carbon Buildings Programme on behalf of the Energy Saving Trust. In order to qualify for LCBP funding, the installation must be carried out by a quality installer, accredited with the Microgeneration Certification Scheme. Further information on the scheme can be found in Chapter 7, and additionally on the internet at www.lowcarbonbuildings.org.uk

Microgeneration Certification Scheme

This was formerly known as the UK Microgeneration Certification Scheme of UKMCS. This is the new Microgeneration Certification Scheme, which has been developed to replace its predecessors, the Clear Skies Accreditation and the Low Carbon Building Programme Accreditation. The scheme, launched to applicants in May 2007, has been developed by the BRE (Building Research Establishment) as a 'robust third party certification scheme', after being commissioned by the then Department of Trade and Industry, which has now been subsumed under the Department for Business, Enterprise and Regulatory Reform.

If you are seeking any form of Government funding for your installation, you will need to ensure that you have sourced the products from manufacturers who have been accredited under the MCS and, furthermore, you will need to ensure that the business you have commissioned to install the products is also certified under the MCS.

As a consumer, you not only want to know that the products you have bought have been manufactured to a sufficiently high standard –

and also installed in a manner that will prove sound and durable – but also that the estimates that you have been given for system performance are accurate and reliable and realistic.

For further information, UK Microgeneration can be contacted at:

Email: ukmicrogeneration@bre.co.uk; tel: 0845 6181514.

Solar Trade Association

Clearly it is in an industry's interest to remove any 'rotten apples' from the barrel, to ensure that the industry maintains its reputation as a whole. The Solar Trade Association (www. greeenergy.org.uk/sta) represents the solar energy industry in the UK, representing manufacturers of equipment as well as installers. Their website allows you to search for members of the STA and they can be contacted for further information on the solar energy Industry in the UK.

Questions To Ask Prospective Installers

Does your company have experience in installing _____?

Different products have different methods of installation, and may require a certain amount of 'learning curve' to install. Familiarizing yourself with a new product – even if you are familiar with the fundamental technology that product is based on – takes time. Fixing methods may differ, electrical connections may be different and performance might vary from one product to the next. Ensure that the installer is not learning on your time! Ask to see evidence of previous uses of your product on other sites.

How many years of experience do you/your company have in installing solar installations?

There is still massive potential for growth in the domestic solar marketplace and, undoubtedly, new companies and individuals will grow to fill the gap in the market and profit from this booming industry. However, there is no substitute for experience and satisfied customers – ask how long your installer has been in the trade, and ask for a list of satisfied customers. Ask in what areas your prospective installer has carried out installations – be wary if they are new to the area and have completed a vast number of installations elsewhere – they may be running away from shoddy jobs!

SUN-PATH DIAGRAMS FOR THE UK

On the following pages is a selection of Sun-path diagrams, which will provide valuable data for those living in the UK. Diagrams have been supplied for 50° latitude (which corresponds with the tip of Cornwall), 55° latitude (which corresponds with Newcastle, Dumfries and Carlisle) and 60° latitude (which corresponds with mainland Shetland). If you live between these lines of latitude, it is possible to view the two graphs adjacent to your latitude, and make a judgement on the position of the Sun, at any time of the year, at any time of the day.

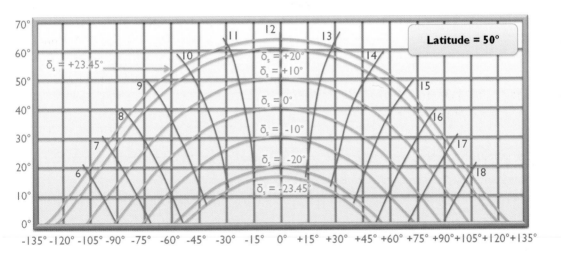

Sun-path diagram for 50° latitude (tip of Cornwall).

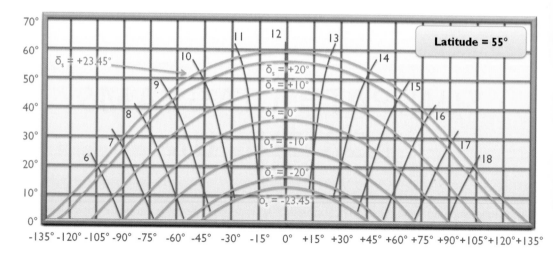

Sun-path diagram for 55° latitude (Newcastle, Dumfries, Carlisle).

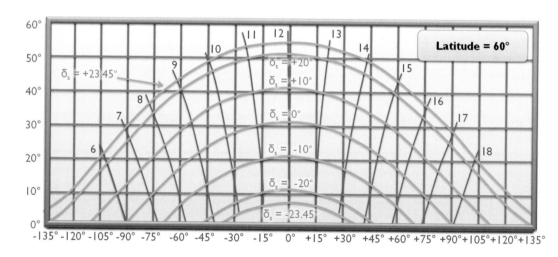

Sun-path diagram for 60° latitude (mainland Shetland).

USEFUL ADDRESSES

Solar Organizations

British Photovoltaic Association
www.greenenergy.org.uk/pvuk2
Now merged with the Renewable Energy
Association.

Solar Trade Association Ltd
Davy Avenue, Knowlhill, Milton Keynes
MK5 8NG
www.greenenergy.org.uk/sta
Email: enquiries@solartradeassociation.org.uk
tel: 01908 442290
fax: 01908 665577

Renewable Energy Association
Formerly Renewable Power Association
www.r-p-a.org.uk

Western Regional Energy Agency and Network
www.wrean.co.uk

Suppliers

Here is an index of all the suppliers of the
products and services featured in this
publication.

Brilliant Energy
3 Tincknells Row, Station Road, Wrington,
North Somerset BS40 5LL

Dulas Ltd
Unit 1, Dyfi Eco Park, Machynlleth, Powys
SY20 8AX
www.dulas.org.uk
Email: webenquiry@dulas.org.uk
tel: 01654 705000
fax: 01654 703000

Invisible Heating Systems
Sustainable Energy Design Centre, Morefield
Industrial Estate, Ullapool
IV26 2SR
www.invisibleheating.co.uk
Email: design@invisibleheating.co.uk
tel: 01854 613 161
fax: 01854 613 16

Mr Pym the Plumber
Datblggu, Penn & Martin Developments Ltd
Brynqwelon, Llanfair Caereinion, Powys
SY21 0BL
www.mr-pym.co.uk
tel: 01938 552020

Rayotec Ltd
Unit 3, Brooklands Close, Sunbury
TW16 7DX
www.rayotec.com/solar_heating
tel: 01932 784848
fax: 01932 784849
Suppliers of SCHOTT solar thermal collectors.

Solar Century
91–94 Lower Marsh, Waterloo, London
SE1 7AB
www.solarcentury.com
Email: enquiries@solarcentury.com
tel: 020 7803 0100
fax: 020 7803 0101

SolarUK
Unit 5, The Estate Yard Buildings, Eridge Green
Road, Tunbridge Wells, Kent TN3 9JR
www.solaruk.net
Email: contactus@solaruk.net ;
tel: 01892 52 63 68

Sol Heat
70 Mostyn Rd, Hazel Grove, Stockport, Cheshire
SK7 5HT
www.solheat.co.uk
Email: customerservices@solheat.co.uk
tel: 0161 612 949
fax: 0871 990 6329

FURTHER READING

The books listed here will provide you with more information on different aspects of utilizing solar energy in your home. Not all of the books focus on an exclusively domestic setting, neither are all written for a UK market, so when conducting further research, be advised that some conventions will differ by country. For US publications, frequency, supply voltage and electrical conventions differ. For books published in the Southern Hemisphere, the direction of the Sun will be reversed and so forth – bear these in mind if information appears contradictory.

General Solar Energy

Martin, Christopher L. and Goswami, D. Yogi (2005) *Solar Energy Pocket Reference.* Earthscan/International Solar Energy Society.

An incredibly handy book, condensing a wealth of solar energy information into a tiny publication, which will fit into your top pocket. Totally invaluable, if you are planning a solar installation.

Schaeffer, John (2004) *Gaiam Real Goods – Solar Living Sourcebook*, 12th edn. New Society Publishers Ltd.

Half book, half catalogue. US bias but an invaluable resource.

Solar Power for Electricity

Daniek, Michel (2007) *Do It Yourself 12 Volt Solar Power: A Do It Yourself Guide.* Permanent Publications.

Introduction to wiring simple 12V DC-only installations – too limited for a domestic installation by far, but ideal for that off-grid shed, summerhouse or retreat in the woods.

German Solar Society – DGS and Ecofys (2007) *Planning and Installing Photovoltaic Systems: A Guide for Installers, Architects and Engineers*, 2nd edn. Earthscan.

Strong, Stephen and Scheller, William (1993) *The Solar Electric House: Energy for the Environmentally-Responsive, Energy-Independent House.* Chelsea Green.

Slightly dated, with a North American focus, but good technical information.

Passive Solar Design

Chriras, Daniel. D. (2004) *The Solar House: Passive Heating and Cooling.* Chelsea Green Publishing Co.

Concentrates on passive solar principles in an

accessible, not-too-technical manner. Written for a US audience, but covers different passive solar strategies for different climates.

Kachadorian, James (2006) *The Passive Solar House: Using Solar Design to Heat and Cool Your Home*. Chelsea Green Publishing Co.

Nice comprehensive, practical book on passive solar design. US bias but more than compensated for by plethora of worksheets and easily readable style.

Van Dresser, Peter (1996) *Passive Solar House Basics*. University of New Mexico Press.

Easily readable book, despite US bias. Nice hand illustrations showing passive solar techniques.

Solar Water Heating

British Standards Institution (1989) *BS 5918:1989 Code of Practice for Solar Heating Systems for Domestic Hot Water*.

This is the British Standard for domestic hot water installations. Contains valuable guidance on good practice in domestic hot water installations.

German Solar Society – DGS and Ecofys (2007) *Planning and Installing Solar Thermal Systems: A Guide for Installers, Architects and Engineers*, 2nd edn. Earthscan.

Despite its cost, this high-quality, colour book is about the most comprehensive guide to planning and installing solar thermal systems available. Technical knowledge assumed.

Laughton, Chris (2006) *Tapping the Sun: A Solar Water Heating Guide*, 5th edn. CAT Publications.

A nice compact easy read covering all the necessary technical detail on solar hot-water heating.

Norton, Brian (2000) *Heating Water by the Sun*, 3rd edn. Solar Energy Society.

Small booklet providing introduction to heating water with the Sun.

Ramlow, Bob and Nusz, Benjamin (2006) *Solar Water Heating: A Comprehensive Guide to Solar Water and Space Heating Systems*. New Society Publishers.

DIY Construction of Solar Collectors

Trimby, Paul (2006) *Solar Water Heating: A DIY Guide*, 5th edn. CAT Publications.

A brief DIY guide from the centre for alternative technology, which focuses on how to construct a solar thermal collector from a radiator. Simple, inexpensive technology, explained clearly and simply.

McCartney, Kevin (1978) *Practical Solar Heating*. Prism.

Although an old source, and out of print, Kevin McCartney's book is still invaluable for the hobbyist looking for a 'solder and flux' practical guide to DIY solar water heating. A likely second-hand internet bargain!

Solar Gadgetry and Experimentation

Harper, Gavin D. J. (2007) *Solar Energy Projects for the Evil Genius*. McGraw Hill.

My own book, US bias, but projects are internationally applicable. Ideal for those who like tinkering in the garage or who want to try and build their own solar-powered gadgetry.

Green Building

Hall, Keith (ed.) (2006) *Green Building Bible: Essential Information to Help You Make Your Home and Buildings Less Harmful to the Environment, the Community and Your Family*. Green Building Press.

One of the best references with an exclusive UK focus with contribution from professionals in all spheres of the green building movement.

The Energy Efficient Home by Patrick Waterfield. Crowood.

Energy Efficiency in Old Houses by Martin Cook. Crowood.

Renewable Energy

Boyle, Godfrey (2004) *Renewable Energy*. Oxford University Press.

If you have a broad-ranging interest in renewable energy and energy systems, I sincerely believe that you would struggle in vain to find a much better introduction than this book. One of the accompanying books to the Open University course T206 – *Energy for A Sustainable Future – Renewable Energy* provides a clear concise, well-illustrated introduction to a range of key concepts.

Renewable Energy: A User's Guide by Andy McCrea. Crowood.

Electrical Installation

Institute of Electrical Engineers/Institution of Engineering and Technology (2008) *Wiring Regulations* (BS7671:2008), 17th edn.

The rulebook for electrical installations – in force for all new electrical installations after 30 June 2008, which must follow the code.

Institute of Electrical Engineers/Institution of Engineering and Technology (2008) *On-Site Guide. Wiring Regulations (BS 7671:2008)*, 17th edn.

This is the accessible face of the 17th edition *Wiring Regulations* – it takes the rules, which are sometimes quite dense and hard to understand, and translates them into practical advice. Not a substitute for the regulations, but certainly very handy to accompany them.

GLOSSARY

Absorber   A dark coloured or selective surface that absorbs the Sun's radiation converting it to heat in the process.

Absorptance   The amount of solar energy absorbed by a surface, relative to the amount of solar energy striking it.

AC   *See* Alternating Current.

Acceptor   Semiconductors are doped 'P' and 'N', by adding a 'dopant' with 'spare' electrons to one piece of semiconductor, whilst doping the other semiconductor with a dopant with 'missing' electrons. An acceptor is a dopant, such as boron, with missing electrons – providing an 'electron hole' to accept electrons.

Active Solar Heating   A solar heating system that uses 'active devices' (e.g. pumps and fans – which move) to move a 'fluid' (e.g. air or water that is being heated), whether for space heating, or for the heating of domestic hot water. This is contrasted with 'passive' solar heating, in which there are no moving parts, and regulation and control of heat is achieved by design, which 'self-regulates' according to the position of the Sun.

Air Mass   The term air mass refers to the amount of atmosphere that solar radiation must pass through before hitting the Earth. For reference, an air mass of 1.0 means the Sun is directly overhead in the sky – travelling through 'one atmosphere' thickness. As the Sun's rays hit the earth obliquely, they must travel through a greater distance of atmosphere, so the air mass increases.

Alternating Current (AC)   A type of electrical current where the direction of flow is reversed on a regular basis. It is convenient to generate alternating current from rotating generators, as a coil cuts through a magnetic flux, and it is easy to transform from a higher to a lower voltage – for this reason it has become the standard in our current electricity generating networks. In the UK, the current changes direction 100 times a second – where each 'cycle' includes two changes of direction. Therefore, there are 50 cycles per second, which we commonly express as a frequency in 'Hertz' – 50Hz.

Altitude   The angle between the horizon and the Sun when facing the orientation of the Sun directly.

Amorphous Semiconductor   Also known as thin-film semiconductor. A semiconductor with no regular order or crystalline lattice.

Amorphous Silicon   *See* Amorphous Semiconductor.

Amp   A unit of electrical current. Electrical current is the flow of electrons in a circuit.

Array   *See* (Photovoltaic) Array.

Auxilliary Heating   The backup to a solar heating installation, delivered by a controllable as opposed to an intermittent energy source, e.g. biomass (preferred) or fossil fuel.

Azimuth   The angle between true south and the point on the horizon below where the Sun appears in the sky.

BERR   *See* Department for Business, Enterprise and Regulatory Reform.

Collector   A device that captures solar energy for transformation into usable energy.

Collector Efficiency   The amount of energy a collector can turn into useful energy, relative to the amount of energy striking it.

Conduction   The flow of heat through a material as a result of the temperature difference between two points in the material. Heat will always travel from a hotter body to a cooler body.

Convection   The flow of heat due to the free movement of particles in a liquid or gas. Hot air rises, because it has a lower density than cold air.

Department for Business, Enterprise and Regulatory Reform   The Government department currently responsible for the UK Renewable Energy Grants Scheme known as the 'Low Carbon Buildings Programme'.

Department of Trade and Industry   Government department, now defunct, under which the remit for renewable energy grants previously fell. The grants now fall under the DTI's successor, BERR (Department for Business, Enterprise and Regulatory Reform).

Direct Insolation   Direct insolation is light that arrives on a solar collecting surface directly from the Sun.

Diffuse Insolation   Diffuse insolation is light that arrives on a solar collecting surface that has been scattered and reflected by clouds, fog or other components of the atmosphere.

DTI   *See* Department of Trade and Industry.

Dual-Axis Tracking   *See* Two-Axis Tracking.

Energy Audit   A detailed survey of energy use in the home, leading to recommendations for areas where energy can be saved.

Equinox   The two times of the year when the night and day are of 'equal' length. Usual dates are 21 March and 23 September.

Evacuated Tubes   Evacuated tubes are a type of solar thermal collector, with a construction similar to a thermos flask, where a glass tube has the air evacuated from it, with a collector surface inside to harness the solar energy.

Fixed Angle Collector   A solar collector whose angle of inclination is fixed and cannot be moved or altered. An example would be a solar collector fixed parallel to a domestic roof or roof integrated solar photovoltaics, where the angle in respect to the Sun does not change or move. (This is the opposite of a tracking array/tracker.)

Flat-Plate Collector A solar collector with the construction of a flat plate collector surface, coated with a material that will absorb the Sun's radiation. The thermal energy generated is taken from the collector by a matrix of pipes through which a working fluid (usually water – air is used on very rare occasions) flows.

Frequency The number of times a waveform repeats itself in a second. In the context of AC power supplies, the UK mains supply is a repeating sine-wave waveform, that repeats fifty times a second. We call this frequency 50 Hertz, expressed as 50Hz.

Grid The National Grid, the infrastructure that connects power stations to consumers of electricity.

Grid-Connected System A solar array whose output is AC power, which is matched and synchronized to the frequency of the National Grid. In a grid-connected system, power can be imported and exported to and from the grid, to balance and compensate for variations in supply and demand.

Grid-Interactive System *See* Grid-Connected System.

Harmonics An AC supply is comprised of a fundamental waveform, which is a pure sine wave. Any devices that connect to an AC circuit, and produce or consume power in a manner that causes current to vary disproportionately with voltage, causes harmonics. Likewise, cheaper inverters that only produce a 'rough' approximation of an AC waveform produce harmonics. Harmonics can present a problem for some loads.

Heat Exchanger A device through which heat is transferred from one fluid to another, without the fluids actually making physical contact. Heat exchangers can be used where it is desirable to transfer heat between two fluids, but where there is a need to keep them physically separated. For example, it is desirable to use antifreeze in the 'solar' portion of a solar domestic hot-water system, but it would be undesirable to have antifreeze coming out of your taps when you go for a wash. A heat exchanger allows the heat from the solar portion of the system to be transferred to the water you use.

Incident Light Light that falls onto a photovoltaic solar cell or collector.

Indirect Insolation *See* Insolation and Diffuse Insolation.

Insolation Insolation is a measure of the power density falling on a surface. Insolation is usually measured in Watts/m². Insolation can be divided into direct insolation and indirect or diffuse insolation. Insolation is a portmanteau of the words 'incoming solar radiation'.

Interconnect A conductor in a solar module/array, which connects solar cell together in either series, parallel or a combination of the two.

Inverter An inverter is a device that takes the DC output from a solar array at a lower voltage than the main's supply and transforms it into a higher voltage AC output. The voltage is increased but the current is decreased. There are a number of different varieties of inverter. Some inverters will generate a stand-alone AC waveform of 50Hz, whilst some inverters will match and synchronize their output to the grid.

Junction (Semiconductor) The region of a semiconductor where the 'N-type' semiconductor and the 'P-type' semiconductors meet. Excitation of free electrons in the semiconductor by photons hitting the junction, cause electrons to 'jump' across this junction, generating a charge.

Kilowatt A unit of electrical power (current × voltage) which is equal to 1,000W.

Kilowatt Hour A power of 1,000W, doing work for a period of one hour.

Life Cycle Cost The cost of owning, running, maintaining and repairing an installation over its useful working life.

Load A device that consumes electrical power.

Maximum Power Point The point on a solar cell's IV curve where the power produced by the cell is at its maximum.

Modified Sine Wave (*See also* Square Wave, Square Wave Inverter) A waveform that is not a perfect sine wave but an approximation, produced by 'steps' of different voltage levels. This strikes a compromise between power quality and the complexity of the circuitry required in the inverter. However, some devices require a 'pure' sine wave to function properly.

Module *See* Photovoltaic Module.

Monocrystalline A type of solar cell that is produced from a single crystal of silicon, which is then sliced up into sheets and turned into solar cells. A characteristic of monocrystalline cells is that they are often round, or have round corners – this results from the process used to produce them.

Multicrystalline *See* Polycrystalline.

Multijunction Device A solar cell that contains more than one 'PN junction' in order to optimize its performance.

Nanometre One billionth of a metre.

Off-Grid System For home-owners living in areas where there is already electricity, using the grid as a 'backup' – somewhere to export extra power and import power when there is a shortfall makes an immense amount of sense. However, solar energy opens up the possibility of delivering electricity to areas without a grid. An off-grid system is one where some back-up is provided in the way of batteries (or maybe in the future hydrogen fuel cells) to store excess energy that is produced when not required, and make that energy available when it is required but not being produced.

Parallel Circuit A parallel circuit is one where all the positives are connected together and all the negatives are connected together. This produces circuits that look like a 'ladder', as opposed to a 'chain'. In a parallel circuit, devices that produce power will produce the same voltage as a single device but with an increased current – and devices that consume power will consume the same voltage as a single device but consume the combined current of all the devices added together.

Photochemical Cell A technology that shows much promise for the future, but which is not yet fully mature and ready for market. Photochemical solar cells produce electrical power as a result of a chemical reaction that is driven by sunlight. The physics underpinning their operation is fundamentally different from photovoltaic cells, and the technology is a distinct and separate one.

(Photovoltaic) Array   A photovoltaic array, PV array, solar array or sometimes just 'array', is a connected system of photovoltaic modules, which together form a single unit, sharing common mounting hardware and structure.

(Photovoltaic) Cell   A photovoltaic cell, PV cell, solar cell or sometimes just 'cell', is the smallest fundamental unit of a solar installation. It consists of a single piece of silicon or substrate coated with thin film. In its raw state, it is not protected against the elements or the outside world. A number of cells are connected together electrically, mounted on a backing substrate and given additional mechanical and environmental protection to form a 'photovoltaic panel'.

(Photovoltaic) Module   A number of photovoltaic cells that are joined together to produce a higher current, voltage or both. Modules are in turn linked together to form panels. Unlike a cell, a photovoltaic module will comprise some form of 'packaging', e.g. cells mounted on a substrate with fixings for mounting hardware, framed by an aluminium strip.

(Photovoltaic) Panel   A photovoltaic panel, PV panel, solar panel, or sometimes just 'panel', refers to a number of photovoltaic modules, which have been connected together electrically and mechanically to form a larger unit.

Polycrystalline   A type of solar cell that is composed of a slice of silicon featuring many silicon crystals, which are a result of the method of construction. Less efficient than monocrystalline solar cells, but perform better than thin-film cells.

Rated Module Current   The current output of a solar module at standard test conditions. *See* Standard Test Conditions.

Semiconductor   A material that neither fully resists, nor fully conducts. The material can be 'doped' with other chemicals to give it exotic electrical properties that allow us to make devices such as photovoltaic cells, which produce power, or devices like transistors and thyristors, which control, manage and amplify power.

Series Circuit   A series circuit is one where devices are connected in a chain, positive to negative, positive to negative, all the way along the chain. In devices that produce power, connecting them in series will increase the voltage produced – but the current will be the same as for one device. For devices that consume power, connecting them in series will increase the voltage consumed – but with the same current that one device would consume.

Sine Wave   The shape of the waveform produced by a plot of voltage over time of the AC power we get from our sockets. Voltage is proportional to the 'sine' of the time. The sine wave is periodic and regular.

Sine-Wave Inverter   An inverter that produces a 'high-quality' sine-wave (AC Power) from a DC source.

(Solar) Array   *See* (Photovoltaic) Array.

(Solar) Cell   *See* (Photovoltaic) Cell.

(Solar) Module   *See* (Photovoltaic) Module.

Solar Noon   The point at which the Sun reaches its highest point in the sky.

(Solar) Panel   *See* (Photovoltaic) Cell.

Solar Resource   The nature, quality and amount of solar radiation that reaches a given site. There are different ways to quantify and measure the solar resource – $kWh/m^2/day$ or peak sun hours.

Solar Time   We use different time systems according to what part of the world we live in. In Britain, we adjust our clocks by one hour for British Summer Time. Solar Time by contrast is the time as determined by the position of the Sun in the sky. It is important when making calculations that take into account the time of day, that we remember to convert between BST, if relevant, and the 'actual' solar time.

Square-Wave   A simple waveform that alternates between two levels. In an AC square wave, the waveform will alternate between a fixed voltage above zero, and a fixed voltage below zero – with an appropriate pause, during which time the supply is at zero to approximate the sine wave waveform of the mains supply.

Square-Wave Inverter   A cheap, basic inverter, that produces an approximation of the AC supply using a simple square wave. (*See also* Modified Sine Wave.)

Staebler–Wronski Effect   The performance of an amorphous silicon solar cell will degrade in the first year of its life and settle down to a lower sunlight to electricity conversion efficiency than when new out of the box. This is due to a degradation process that occurs upon first exposure to light, and is as a result of the Staebler–Wronski effect.

Stand-Alone System   *See* Off-Grid System.

Standard Test Conditions   Standard test conditions are used to make an objective comparison between different solar modules of different solar manufacturers, so that a reliable judgement on the relative merits of different solutions can be made. A cell being exposed to standard test conditions will be at an ambient temperature of $25°C$ whilst being exposed to an irradiance of $100W/m^2$.

String   A number of solar cells connected in series to produce a desired voltage. A 'string' of cells can be connected to a 'string inverter'.

String Inverter   An inverter designed to be wired to a single series string of a number of solar cells.

Substrate   A material that provides mechanical strength or backing to a solar cell, or a base on which a film of material can be coated.

Thermal Mass   Materials that have the ability to store thermal energy. Materials with high thermal mass can be used to store thermal energy, or even-out temperature variations over time, by releasing heat slowly.

Thin-Film   A solar device constructed of thin layers of semiconducting material – either exotic metal oxides or amorphous silicon. The share of thin-film solar cells is expected to grow rapidly, as although less efficient than crystalline solar cells, they have the potential to be produced much more cheaply. Thin film solar cells can also be coated onto a flexible substrate, which opens up the possibility of 'bendy' solar cells, opening doors for architectural creativity. *See also* Amorphous Semiconductor.

Tracker/Tracking Array/Solar Tracker A solar array (thermal or electrical) that changes its position in order to follow the path of the Sun. This enables the maximum amount of solar radiation to reach the solar array. There are two methods – single and two (or dual) axis tracking. The first is the simplest – the array is set at a fixed tilt-angle, and the tracker follows the Sun from east to west as it rises and sets. In two axis tracking, the array changes tilt-angle (inclination) in addition to following the Sun from east to west.

Tilt Angle/Inclination The angle to which a solar array or collector is inclined with respect to the ground.

Two-Axis Tracking A tracking solar array, which can track the sun in both its altitude and azimuth – following the path of the Sun over the course of a day, and adjusting for the changing height of the Sun – as well as changing the direction of the solar array to face the Sun.

Ultraviolet The component of the Sun's light that is beyond the violet end of the visible spectrum, and occupies the wavelength range of 4 –400 nanometres (nm).

Volt (V) The volt is a unit of voltage. One volt when acting on a resistance of one ohm produces a current of one ampere.

Voltage Voltage is a measurement of the 'force' of electrons in an electrical circuit. We can think of voltage as being equivalent to pressure in a pipe.

Volt Drop/Voltage Drop The amount of voltage lost along the length of a conductor because of the conductor's resistance. We can calculate the volt-drop by using Ohm's Law.

Wafer A thin slice of semiconductor, which can be produced in a number of ways. It is the raw material from which crystalline solar cells are produced.

Watt (W) In a DC circuit, one amp of current at a potential of one volt is equal to one watt of power. The relationship between current, voltage and power becomes more complicated when looking at AC power.

Watt-Hour (Wh) One Watt-hour, is equivalent to a unit of electrical power equivalent to one watt, flowing for a duration of one hour.

Zenith The zenith is the point directly above you, which is perpendicular to the Earth's surface. If you were standing on the Equator during the Equinox at solar noon, the Sun will be at the Zenith of the sky.

Zenith Angle The Zenith angle is the angle between the Zenith of the sky and the object (usually the Sun) that we want to measure.

INDEX